# Two Manuals Combined into One Book



# Light Antiarmor Weapons

# And

# Army M47 Dragon Med Antitank Weapon

Department of Defense

Pentagon
Publishing

FIELD MANUAL
NO. 3-23.25 (FM 23-25)

HEADQUARTERS
DEPARTMENT OF THE ARMY
WASHINGTON, DC, 30 August 2001

# LIGHT ANTIARMOR WEAPONS

## CONTENTS



DISTRIBUTION RESTRICTION: Approved for public release; distribution is unlimited.

*This publication supersedes FM 23-25, 17 August 1994.

# PREFACE

This manual provides technical information, training techniques, and combat techniques of light antiarmor weapons. Intended users include leaders and designated soldiers who will use this information to successfully integrate light antiarmor weapons into combat operations. This manual discusses gunnery training and train-the-trainer.

The tactical positions shown in this manual were drawn to enhance the reader's understanding of related subject material and do not necessarily represent the best employment option for that material.

Unless this publication states otherwise, masculine nouns and pronouns do not refer exclusively to men.

The proponent of this publication is the United States Army Infantry School. Send comments and recommendations on DA Form 2028 (Recommended Changes to Publications and Blank Forms) directly to—

Commandant
US Army Infantry School
ATTN: ATSH-IN-S3
Fort Benning, GA 31905-5595
lusanoh@benning.army.mil

CHAPTER 1

# INTRODUCTION

*This chapter provides information common to the light antiarmor weapons discussed in this manual. Topics include care and handling, destruction and decontamination procedures, and operating temperatures. Light antiarmor weapons are used against light armored vehicles, field fortifications, or other similar targets. These weapons are issued as rounds of ammunition to individual soldiers in addition to their assigned weapons and the unit's organic antiarmor weapons. Light antiarmor weapons can withstand extreme weather and environmental conditions, including arctic, tropical, and desert. The light antiarmor weapons category includes both light antiarmor and light antitank weapons.*

## 1-1. TYPES OF LIGHT ANTIARMOR WEAPONS

Light antiarmor weapons include the M72-series light antitank weapon (LAW) and the M136 AT4. The M72-series LAW was designed in the early 1960's for use against light tanks of that era (Figure 1-1). The M136 AT4 was designed in the late 1980's for use against the improved armor of light armored vehicles (Figure 1-2).

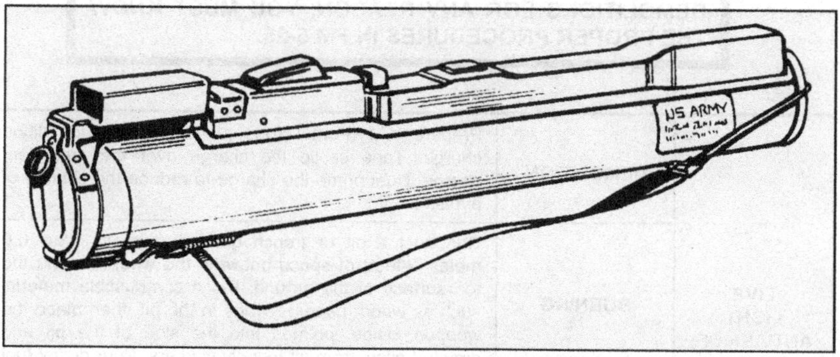

**Figure 1-1. M72-series light antitank weapon.**

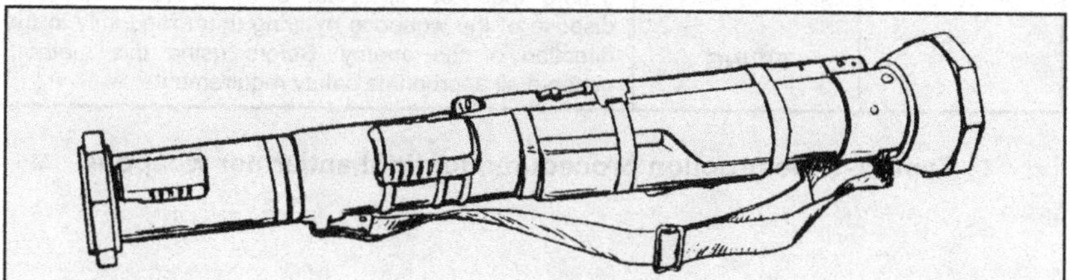

**Figure 1-2. M136 AT4 light antiarmor weapon.**

## 1-2. CARE AND HANDLING

Light antiarmor weapons are issued as rounds of ammunition. The only requirement for their care is a visual inspection, outlined in the appropriate chapter for each weapon (Chapter 2 for the LAW and Chapter 3 for the AT4).

## 1-3. DESTRUCTION PROCEDURES (COMBAT ONLY)

In combat, live and expended light antiarmor weapons are destroyed only to prevent their capture or use by the enemy and, even then, only on order. For such an order to be given, the weapons must be so badly damaged that neither repairs nor cannibalization can restore them to usable condition (FM 5-25). Table 1-1 provides destruction procedures for live and expended light antiarmor weapons; Appendix A discusses safety precautions to follow when destroying them.

**DANGER**

**TO AVOID POSSIBLE INJURY OR DEATH, MOVE TO A SAFE POSITION AND TAKE COVER BEFORE USING ANY DESTRUCTION PROCEDURE. BEFORE USING DEMOLITIONS FOR ANY REASON, YOU MUST KNOW THE PROPER PROCEDURES IN FM 5-25.**

| | | |
|---|---|---|
| **LIVE LIGHT ANTIARMOR ROUND** | **DEMOLITION** | Prepare a 113-gram (one-quarter pound) demolition charge. Tape or tie the charge over the propellant charge. Dual prime the charge to reduce the chance of a misfire. |
| | **BURNING** | Construct a pit or trench deep enough to allow 0.6 meter (2 feet) of space between the weapons and the top surface of the ground. Place combustible material such as wood, paper, or rags in the pit, then place the weapon inside, pointed into the side of the pit and directed away from all friendly soldiers. Pour diesel fuel or oil over the weapons and the combustible material. |
| | **FIRING** | If time does not permit use of the previous methods, dispose of the weapons by firing them randomly in the direction of the enemy. Before using this method, observe all appropriate safety requirements. |

**Table 1-1. Destruction procedures for light antiarmor weapons.**

| | | |
|---|---|---|
| **EXPENDED LIGHT ANTIARMOR ROUND** | **MECHANICAL** | Though you must never use mechanical means to destroy live antiarmor weapons or expended M136 AT4s, you may do so to destroy the residue from an expended M72-series LAW. For example, you can drive over it with a tracked vehicle or strike it with a pick, ax, or other object, as long as you make it unusable. |
| | **DEMOLITION** | Same as a live round. |
| | **BURNING** | Same as a live round. |

**Table 1-1. Destruction procedures for light antiarmor weapons (continued).**

**DANGER**

**1. WHEN USING FIRE TO DESTROY A LIGHT ANTIARMOR WEAPON, THE TIME REQUIRED TO EXPLODE THE WARHEAD IS UNPREDICTABLE. ALSO, IGNITING THE PROPELLANT CAN CAUSE IT TO FIRE THE WARHEAD IN ANY DIRECTION, WHICH COULD IN TURN CAUSE INJURY OR DEATH.**

**2. OBSERVE THE APPROPRIATE SAFETY PRECAUTIONS WHEN HANDLING DIESEL FUEL. CARELESSNESS COULD CAUSE PAINFUL, EVEN FATAL, BURNS.**

**3. DO NOT TRY TO USE VEHICLES OR MECHANICAL MEANS TO DESTROY LIVE ANTIARMOR WEAPONS. EITHER METHOD COULD DETONATE THE WARHEAD OR PROPELLANT CHARGE, WHICH COULD CAUSE INJURY OR DEATH.**

## 1-4. DECONTAMINATION PROCEDURES

The soldier can use his M258A1 or DKIE (XM280) individual decontamination packet to remove H-series, G-series, and V-series agents. FM 3-5 provides more information about decontamination procedures for equipment and weapons.

**DANGER**

**NEVER USE DS2 TO DECONTAMINATE ANY LIGHT ANTIARMOR WEAPON. THE DS2 WOULD DISSOLVE THE RUBBER AND PLASTIC SEALS, ALLOWING THE DS2 TO REACH THE PROPELLANT AND PRODUCING AN EXTREMELY HAZARDOUS MIXTURE.**

## 1-5. OPERATING TEMPERATURES

Operating temperatures for the M72-series LAW and M136 AT4 are -40 degrees to 140 degrees Fahrenheit (-40 degrees to 60 degrees Centigrade). Firing light antiarmor weapons in temperatures outside these limits could cause a misfire or produce some other hazard for the soldier (Appendix A).

## CHAPTER 2
# M72-SERIES LAW, OPERATION AND FUNCTION

*This chapter provides information on and technical data for the M72-series light antitank weapon (LAW). It also discusses the characteristics, nomenclature, functioning, and operation of the LAW.*

### 2-1. DESCRIPTION

The M72-series LAW is a lightweight, self-contained, antiarmor weapon consisting of a rocket packed in a launcher (Figure 2-1). It is man-portable, may be fired from either shoulder, and is issued as a round of ammunition. It requires little from the user—only a visual inspection and some operator maintenance. The launcher, which consists of two tubes, one inside the other, serves as a watertight packing container for the rocket and houses a percussion-type firing mechanism that activates the rocket.

a. **Outer Tube**. The trigger housing assembly (which contains the trigger assembly) is on the upper surface of the outer tube. So are the trigger arming handle, front and rear sight assemblies, and the launcher's rear cover.

b. **Inner Tube**. The inner tube telescopes outward toward the rear, guided by a channel assembly that rides in an alignment slot in the outer tube's trigger housing assembly. The channel assembly also houses the firing pin rod assembly, which includes a detent lever assembly. The detent lever assembly moves under the trigger assembly in the outer tube, locking the inner tube in the extended position and cocking the weapon. All this must occur before the weapon can be fired.

c. **Rocket**. The rocket is a percussion-ignited, fin-stabilized, fixed munition. The rocket is attached by the igniter to the inside of the launcher. The rocket consists of a 66-mm HEAT warhead, a point-initiating, base-detonating fuze, and a rocket motor. Six spring-loaded fins are attached to the rear of the rocket motor. These fins are folded forward along the motor when the rocket is in the launcher. When ignited, the propellant in the rocket motor burns completely, producing gasses about 1,400 degrees Fahrenheit (760 degrees Centigrade). The gas pressure pushes the rocket toward the target and exits to the rear of the launcher as the backblast.

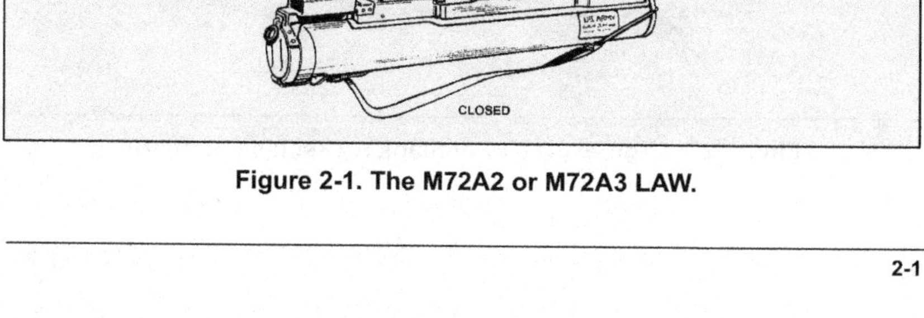


**Figure 2-1. The M72A2 or M72A3 LAW.**

## 2-2. TECHNICAL DATA

The following data apply to the M72A2 and M72A3 LAWs:

a. **Launcher.**

Length (Extended)....................Less than 1 meter (34.67 inches)
Length (Closed)....................0.67 meter (24.8 inches)
Weight (Complete M72A2)....................2.3 kg (5.1 pounds)
Weight (Complete M72A3)....................2.5 kg (5.5 pounds)
Firing Mechanism....................Percussion
Front Sight....................Reticle graduated in 25-meter range increments
Rear Sight....................Peep sight adjusts automatically to temperature change

b. **Rocket.**

Caliber....................66 mm
Length....................50.8 cm (20 inches)
Weight....................1.8 kg (2.2 pounds)
Muzzle Velocity....................144.8 mps (475 fps)
Minimum Range (Combat)....................10 meters (33 feet)
Minimum Arming Range....................10 meters (33 feet)
Maximum Range....................1,000 meters (3,300 feet)
*Maximum Effective Ranges*:
Stationary Target....................200 meters (660 feet)
Moving Target....................165 meters (541 feet)
(Beyond these ranges, there is less than a fifty percent chance of hitting the target.)

## 2-3. AMMUNITION

The M72-series LAW is issued as a round of ammunition. It contains a nonadjustable propelling charge and a rocket. Every M72-series LAW has an integral high-explosive antitank (HEAT) warhead. The warhead is in the rocket's head (or body) section. The fuze and booster are in the rocket's closure section. The propellant, its igniter, and the fin assembly are in the rocket's motor. No inert versions are available (Figure 2-2). Appendix B provides information about appropriate gunnery training devices and ammunition. Although the M72-series LAW is mainly used as an antiarmor weapon, it may be used with limited success against secondary targets such as gun emplacements, pillboxes, buildings, or light vehicles. (Chapter 6 provides more information about combat techniques.)

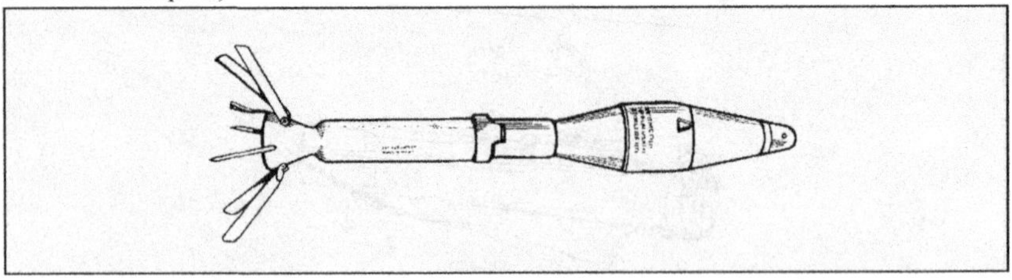

**Figure 2-2. High-explosive antitank rocket (ammunition).**

a. **Description**. The 66-mm HEAT rocket warhead consists of a tapered, thin-gauge steel body. When it explodes, the force and heat of the explosive focus into a small but powerful gas jet. This directional jet penetrates the target and, if the target is a vehicle, sprays molten metal inside. If the jet hits an engine or ammunition, it may start a fire or cause an explosion. Figure 2-3 shows how the warhead penetrates 300 millimeters of rolled homogeneous steel armor.

(1) ***Impact***. The nose cone crushes; the impact sensor activates the fuze.

(2) ***Ignition***. The ogive crush switch activates the electric detonator. The booster detonates, initiating the main charge.

(3) ***Penetration***. The main charge fires and forces the warhead body liner into a directional gas jet that penetrates armor plate.

(4) ***After Armor Effects (Spalling)***. The projectile fragments and incendiary effects produce blinding light and destroy the target's interior.

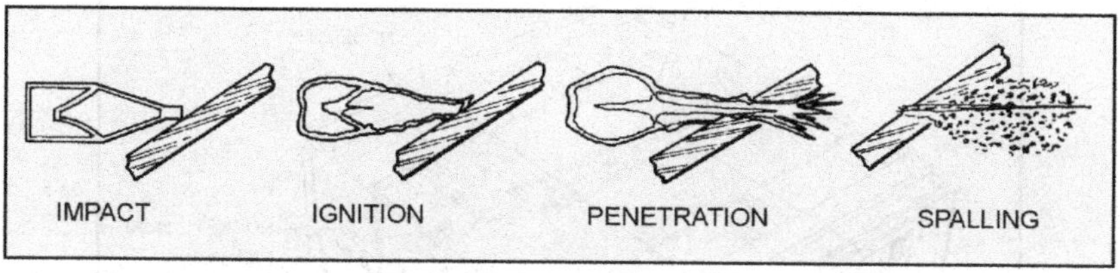


**Figure 2-3. Effects of M72-series LAW warhead.**

b. **Characteristics**. The head of the round is olive drab stenciled in yellow. The M412 fuze is dropsafe and boresafe. Its minimum arming distance is about 33 feet (10 meters). Six stabilizing fins are attached as part of the motor. As the rocket clears the launcher, springs force open the fins, which stabilize the rocket in flight.

c. **Packaging**. Five complete M72-series LAWs are packaged within a fiberboard inner pack for a total weight of 12.5 kilograms (27 1/2 pounds). Three inner packs are then placed in a wire-bound wooden box, the gross weight of which is 54.5 kilograms (120 pounds) (Figure 2-4, page 2-4).

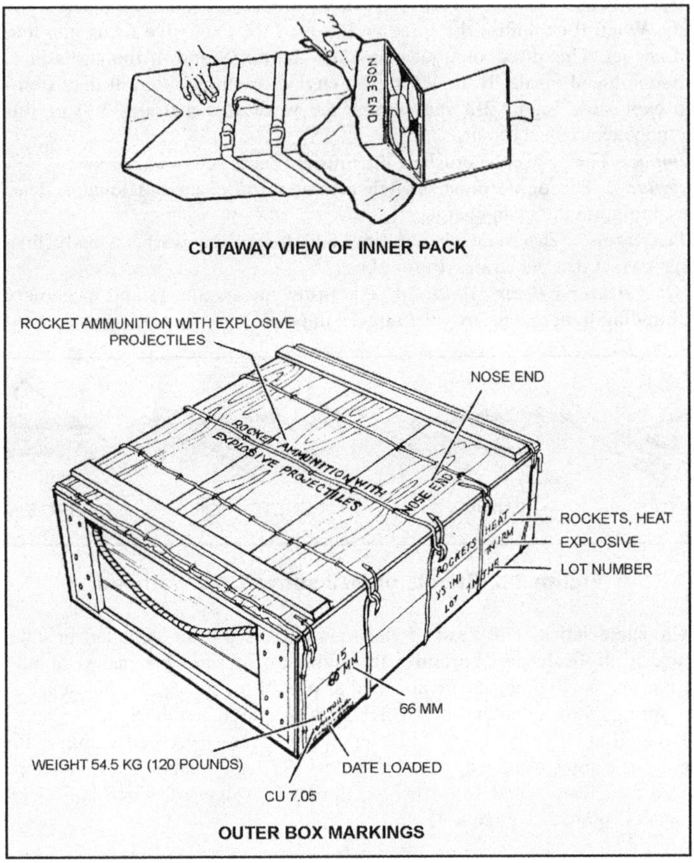


**Figure 2-4. Packaging for M72-series LAW.**

## 2-4. INSPECTION

Because the M72-series LAW is issued as a round of ammunition rather than as a weapon, inspection is limited to a visual examination of the sealed unit.

Inspect the launcher's overall condition before preparing the launcher for use.

- Check the body for dents, cracks, or bulges.
- Check the rubber boots covering the trigger bar and barrel detent for tears or punctures.
- Ensure the arming handle is present and on SAFE and that the pull pin is in place.

- Check the data plate for the phrase, W/COUPLER (Figure 2-5).

**DANGER**

**1. IF THE M72A2 LAW DOES NOT STATE "W/COUPLER" ON ITS DATA PLATE, TURN THE WEAPON IN TO THE UNIT AMMUNITION SECTION.**

**2. THE COUPLER PREVENTS THE INNER AND OUTER TUBES FROM SEPARATING AND POSSIBLY CAUSING PREMATURE DETONATION.**

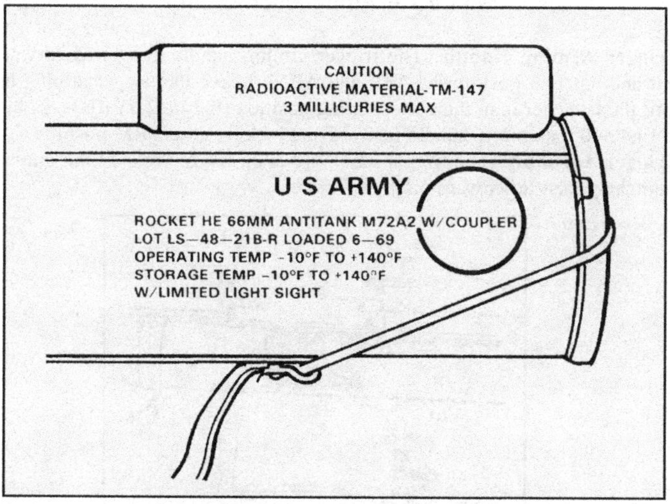


**Figure 2-5. Launcher data plate**

## 2-5. FIRING MECHANISM

The firing mechanism includes the trigger arming handle, the trigger assembly, and the firing pin rod assembly (Figure 2-6).

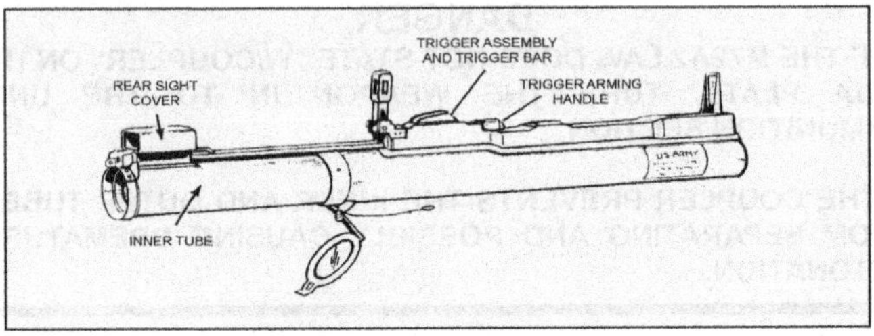


**Figure 2-6. Firing mechanism.**

a. **Trigger Arming Handle**. The trigger arming handle is located forward of the trigger bar and has two positions: SAFE and ARM. Leave the trigger arming handle on SAFE until the launcher is in the correct firing position (Figure 2-7). To press the trigger, you must first pull the arming handle forward and lock it in the ARM position.

b. **Trigger Assembly**. The trigger assembly is on the top rear of the outer tube. To fire the launcher, press downward on the trigger bar.

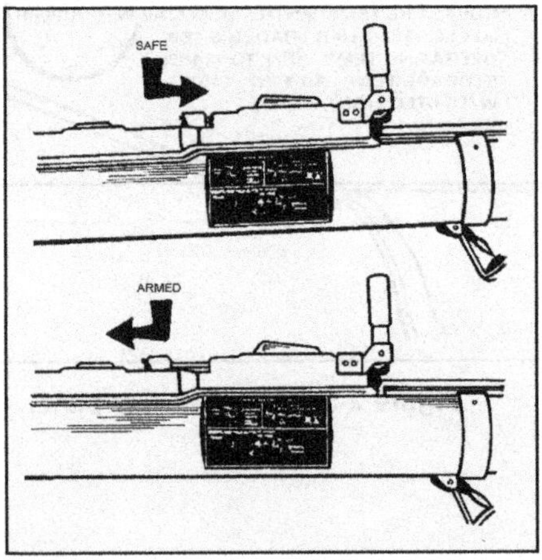


**Figure 2-7. Trigger arming handle.**

c. **Firing Pin Rod Assembly**. The rear sight cover and the firing pin housing are on the top of the rear of the inner tube. Inside the housing, the primer and the firing pin rod are aligned (Figure 2-8). Pressing the trigger bar releases the tension on the firing pin rod assembly, allowing the firing pin to strike the center of the primer.

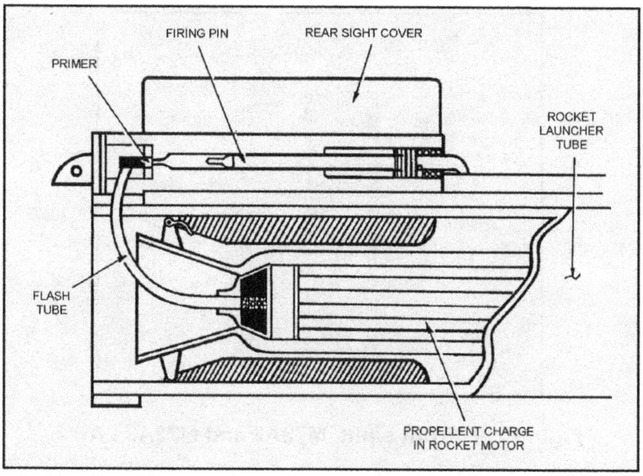


**Figure 2-8. Firing pin, primer, and propellant charge.**

## 2-6. SIGHTS

This paragraph discusses the front and rear sights and their proper use.

a. **Front Sight, M72A2 and M72A3 LAWs.** The front sight has a raised vertical range line marked with ranges from 50 to 350 meters in 25-meter increments (Figure 2-9). Two curved stadia lines are etched on the front sights. Do not use the stadia lines on this sight to estimate range, because they are inaccurate. Lead indicators are located on either side of the stadia lines to help you engage moving targets. On the M72A3, use the front sight illuminated range marks at the 100-meter and 150-meter points to help you engage targets in low light.

**WARNING**

**DO NOT TOUCH THE RANGE MARKS; THEY ARE ILLUMINATED WITH PROMETHIUM, WHICH IS MILDLY RADIOACTIVE.**

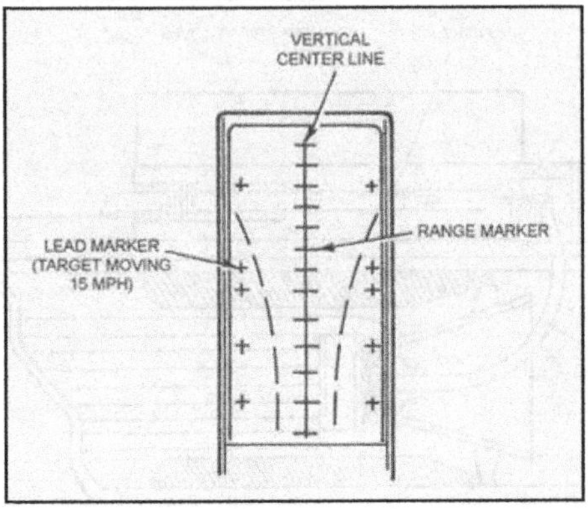


**Figure 2-9. Front sight, M72A2 and M72A3 LAWs.**

b. **Rear Sight, M72A2 and M72A3 LAWs**. The rear sight consists of a steel bracket with a rubber boot and plastic peep sight. This sight automatically adjusts to changes in temperature (Figure 2-10), which means that its settings are unaffected by temperature.

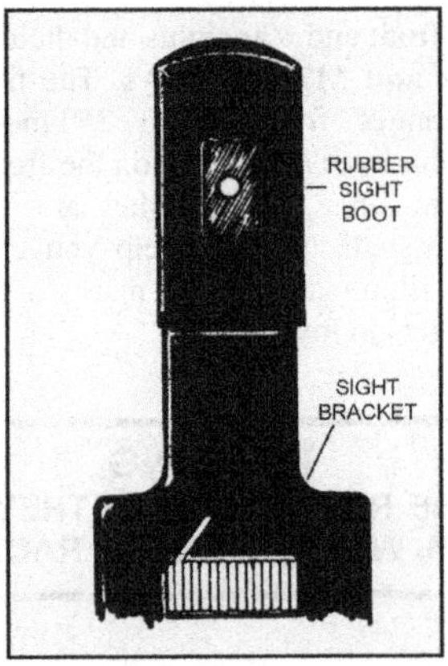


**Figure 2-10. Rear sight, M72A2 and M72A3 LAWs.**

c. **AN/PVS-4 Nightsight**. This nightsight is issued with various accessories, including a bracket that, when mounted on an M72-series LAW, will allow you to use an AN/PVS-4 on the LAW. However, this works only if DS maintenance has already installed an M72A1 reticle in the AN/PVS-4. Though the reticle was developed for the M72A1 LAW, you can also use it with the other M72-series models. To mount the M72A1 bracket assembly on any M72-series LAWC

(1) Place the bracket assembly on top of the rocket launcher (Figure 2-11) so that the square cutout in the top of the bracket fits over the extension release button.

(2) Swing the lower adapter section up and under the rocket launcher and secure it by turning the locking latch clockwise to fully engage the latch shoulder screw.

(3) Place the sight in the groove on the bracket and align the threaded screw hole in the base of the sight with the lever screw assembly. Tighten the lever screw assembly firmly.

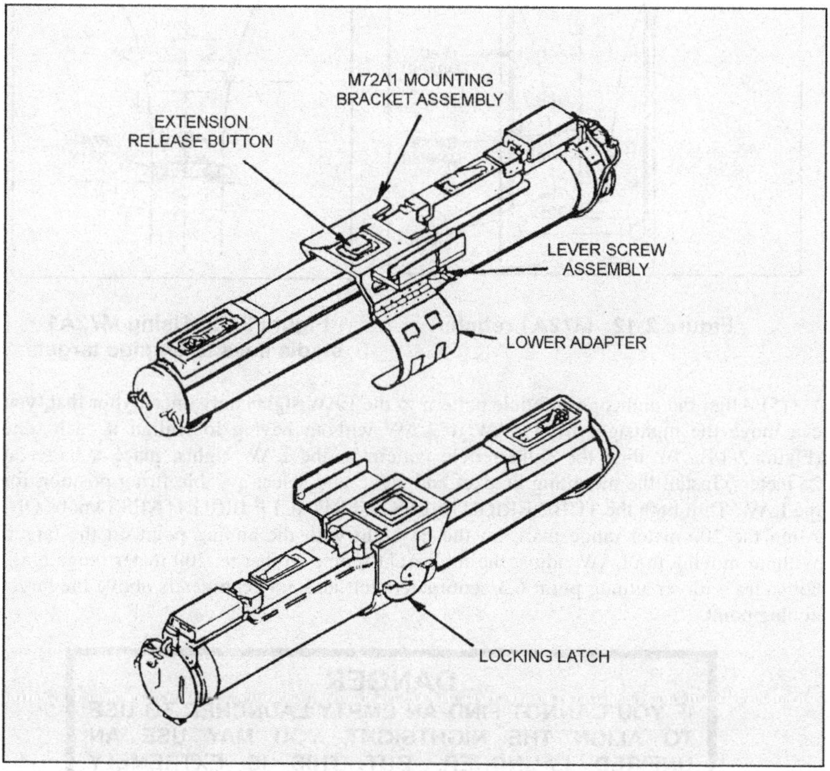


**Figure 2-11. Installing M72A1 mounting bracket assembly.**

(4) Use the M72A1 reticle to estimate the range to a 20-foot target such as a tank viewed from the side (Figure 2-12). Place the sight so that the rear and front of a flanking target fit between the curved vertical lines (Figure 2-13). When you read the range from the scale, note that the width of the tank is about one-half the tank's length. This means you can estimate the range to a target, whether it is headed straight toward you or straight away from you, by placing its sides between the curved vertical lines and halving the range shown at that point. The bottom of each vertical line in the center of the reticle

corresponds to an additional 25 meters. The length of a horizontal line represents 5 mils, and the spaces between lines represent 5 mils. Use the stadia lines to estimate range *only* when using the AN/PVS-4's M72A1 reticle pattern.

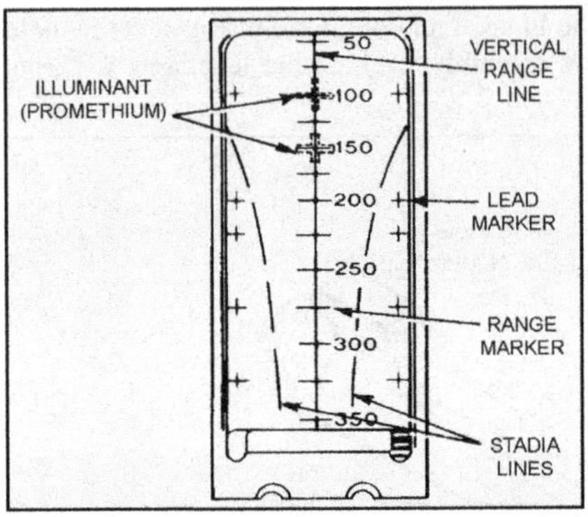


**Figure 2-12. M72A1 reticle.**

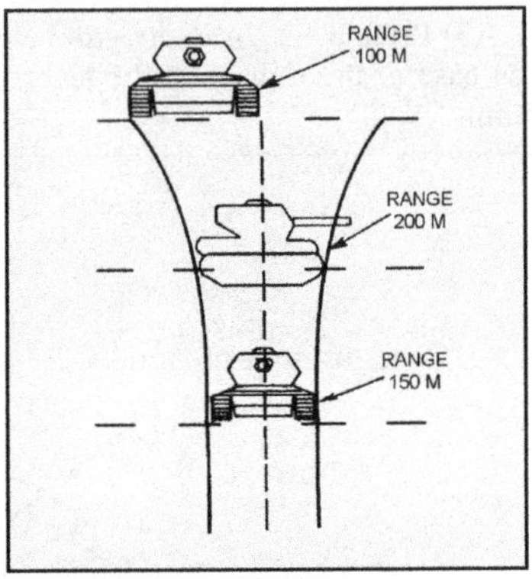


**Figure 2-13. Using M72A1 stadia lines to engage targets.**

(5) Align the nightsight's reticle pattern to the LAW sights only once. After that, you can move the nightsight from LAW to LAW without having to realign it each time (Figure 2-14). To align the sight reticle pattern to the LAW sights, place a target at 25 meters. Install the mounting bracket and sight, and select a stable firing position for the LAW. Turn both the TUBE BRIGHTNESS and RETICLE BRIGHTNESS knobs ON. Align the 200-meter range mark on the daysight with the aiming point on the target. Without moving the LAW, adjust the nightsight reticle so that its 200-meter range mark coincides with an aiming point 6.3 centimeters left and 3.5 centimeters above the target aiming point.

**DANGER**

**IF YOU CANNOT FIND AN EMPTY LAUNCHER TO USE TO ALIGN THE NIGHTSIGHT, YOU MAY USE AN UNFIRED LAUNCHER, BUT THIS IS EXTREMELY DANGEROUS.**

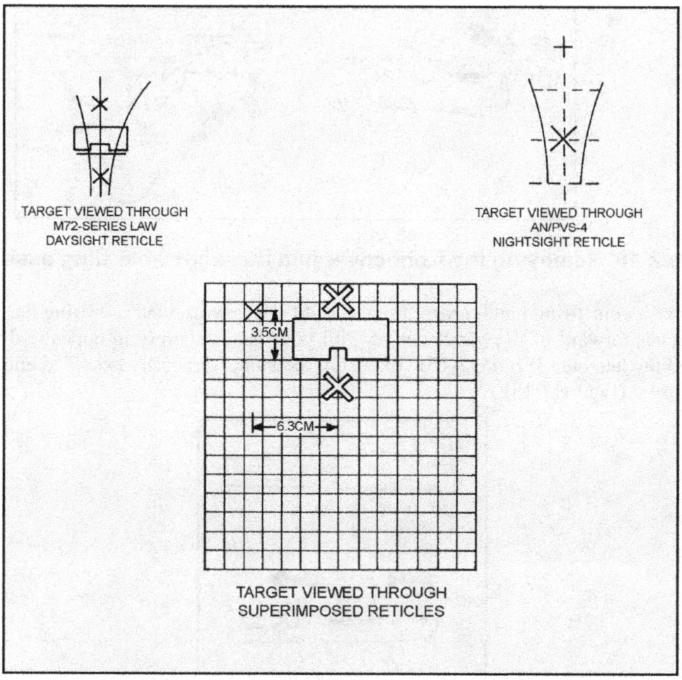


**Figure 2-14. M72A1 alignment target for the AN/PVS-4.**

## 2-7. OPERATION AND FUNCTION

Before preparing the launcher for use, the firer inspects its overall condition.

a. To extend the rocket launcher—

(1) Remove the pull pin and rotate the rear cover downward so the front cover and adjustable sling assembly can fall free (Figures 2-15 and 2-16, page 2-11). Do not discard the sling assembly until after you fire the rocket.

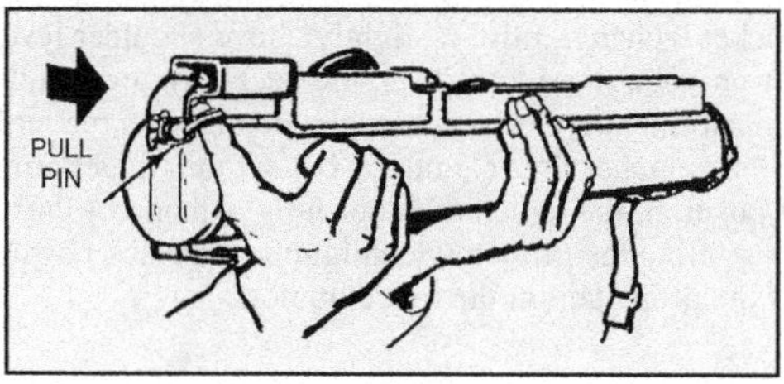


**Figure 2-15. Removing the pull pin.**

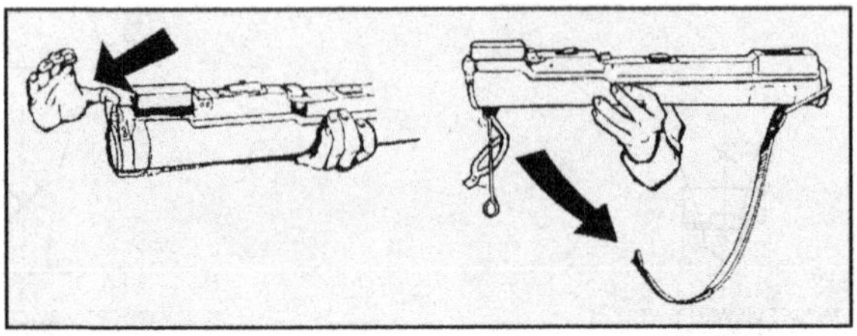

**Figure 2-16. Removing the front cover and the adjustable sling assembly.**

(2) With your firing hand, grasp the rear sight cover; with your nonfiring hand, grasp the launcher forward of the barrel detent. Pull your hands sharply in opposite directions to extend the launcher (Figure 2-17). To ensure the launcher is fully extended and locked, try to close it (Figure 2-18).

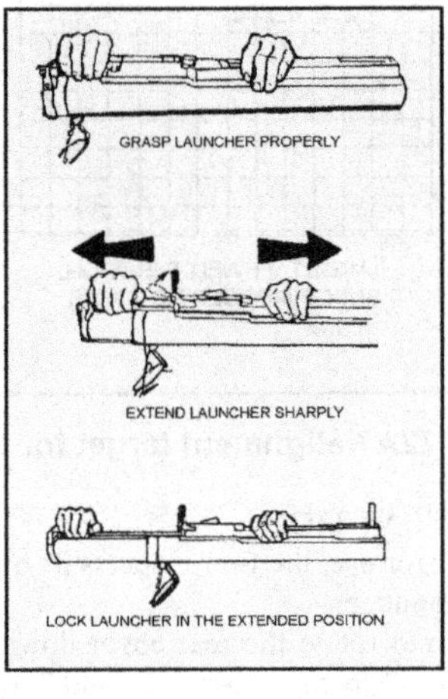


**Figure 2-17. Extending the launcher.**

b. To fire the rocket launcher, raise it slightly above shoulder level, rotate your body under it, and place it on your shoulder. Check the backblast area, pull the trigger arming handle to the ARM position, aim the launcher, and depress the rubber boot on the trigger bar firmly to ensure the launcher fires (Figure 2-19). (If the trigger arming handle will not remain in the ARM position, the launcher is not fully extended.) Pressing the trigger bar causes the firing pin to strike the primer, which ignites the black powder in the flash tube, which in turn ignites the propellant in the rocket motor.

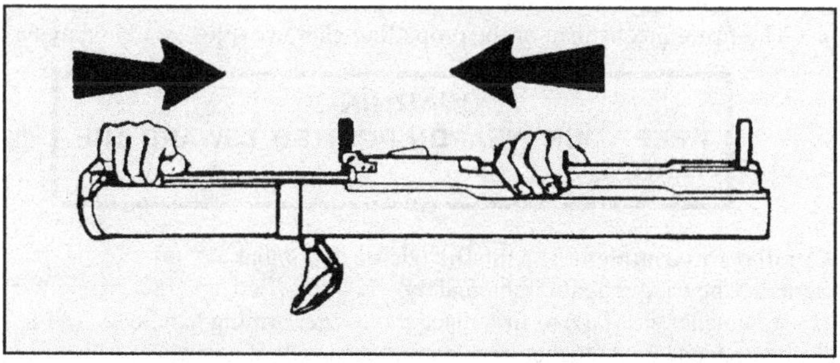

**Figure 2-18. Ensuring launcher is locked in the extended position.**

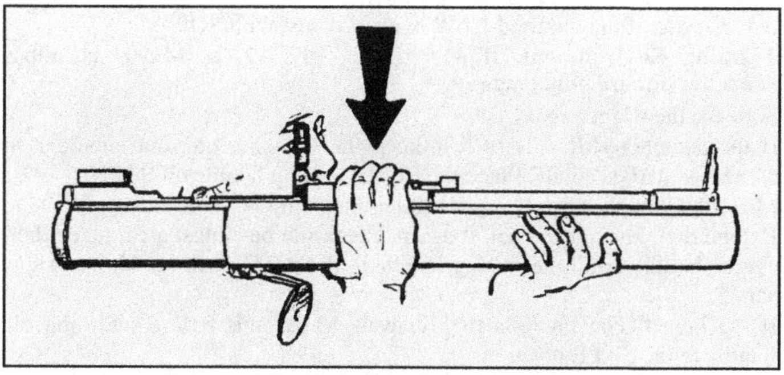

**Figure 2-19. Firing the launcher.**

> **DANGER**
> **WHEN OPERATING THE LAW, KEEP IT POINTED DOWNRANGE. ENSURE YOUR WHOLE BODY IS CLEAR OF THE MUZZLE AND REAR OF THE LAUNCHER, AND ENSURE THE BACKBLAST AREA IS CLEAR.**

## 2-8. MISFIRE PROCEDURES

A misfire is a complete failure to fire caused by a procedural or mechanical failure. Which misfire procedures should be used depends on whether the firer is in a combat or training environment.

a. **Causes**. A misfire is usually caused by one of the following factors:

- The launcher may not be fully extended.

- The trigger arming handle may not be armed.
- The firing mechanism or the propelling charge explosive train may be faulty.

> **WARNING**
> **KEEP YOUR WEAPON POINTED TOWARD THE TARGET.**

b. **Combat Environment**. If a misfire occurs in combat—

(1) Squeeze the trigger again immediately.

(2) If the launcher still fails to fire, place the trigger arming handle on SAFE.

(3) Partly collapse the launcher, then extend it to cock it again.

(4) Place it on your shoulder, check the backblast area again, then arm, aim, and fire.

(5) If the LAW still fails to fire, squeeze the trigger again and return the trigger arming handle to SAFE. Collapse the launcher, set it aside, and try another one. As soon as you can, dispose of the misfired LAW in accordance with SOP.

c. **Training Environment**. If an M72A2, M72A3, or M190 subcaliber device misfires on a live-fire training range—

(1) Squeeze the trigger again.

(2) If the launcher still fails to fire, keep the launcher on your shoulder, announce "Misfire," and wait 10 seconds. Place the trigger arming handle on SAFE.

(3) Move the launcher from your shoulder and wait one minute.

(4) Extend the launcher to cock it again, check the backblast area, place the launcher back on your shoulder, pull the arming handle to the ARM position, aim, and squeeze the trigger bar.

(5) If the launcher again fails to fire, wait 10 seconds before returning the trigger arming handle to the SAFE position.

(6) Keep the launcher trained on the target area at least one minute; DO NOT collapse the launcher.

(7) Move the launcher to a safe area and dispose of it IAW unit SOP.

## 2-9. RESTORATION TO CARRYING CONFIGURATION

If the launcher is prepared to fire, but then is not fired, it should be returned to the carrying configuration by reversing the preparation procedure. After the launcher has been prepared for firing, it is no longer watertight. Therefore, when carrying the launcher, sling it over either shoulder with the muzzle (forward) end down. Only the rocket and rocket motor ignition system are waterproof.

a. Return the trigger arming handle to the SAFE position.

b. Remove the launcher from your shoulder, depress the barrel detent, collapse the launcher tube, and guide the front and rear sights into position.

c. Close the rear cover, replace the cover pull pin, and replace the sling assembly.

> **WARNING**
> **TO PREVENT INJURY, REMOVE YOUR THUMB FROM THE DETENT AFTER COLLAPSING THE LAUNCHER 1/2 TO 1 INCH.**

CHAPTER 3

# M136 AT4, OPERATION AND FUNCTION

*This chapter provides information and technical data for the M136 AT4 light antiarmor weapon, including its characteristics, nomenclature, and operation. Its function, firing mechanism, and safeties are also discussed.*

## 3-1. DESCRIPTION

The M136 AT4 is a lightweight, self-contained, antiarmor weapon. It consists of a free-flight, fin-stabilized, rocket-type cartridge packed in an expendable, one-piece, fiberglass-wrapped tube (Figure 3-1). The M136 AT4 is man-portable and is fired from the right shoulder only. The launcher is watertight for ease of transportation and storage. Though the M136 AT4 can be employed in limited visibility, the firer must be able to see and identify the target and estimate the range to it. Unlike the M72-series LAW, the M136 AT4 launcher need not be extended before firing.

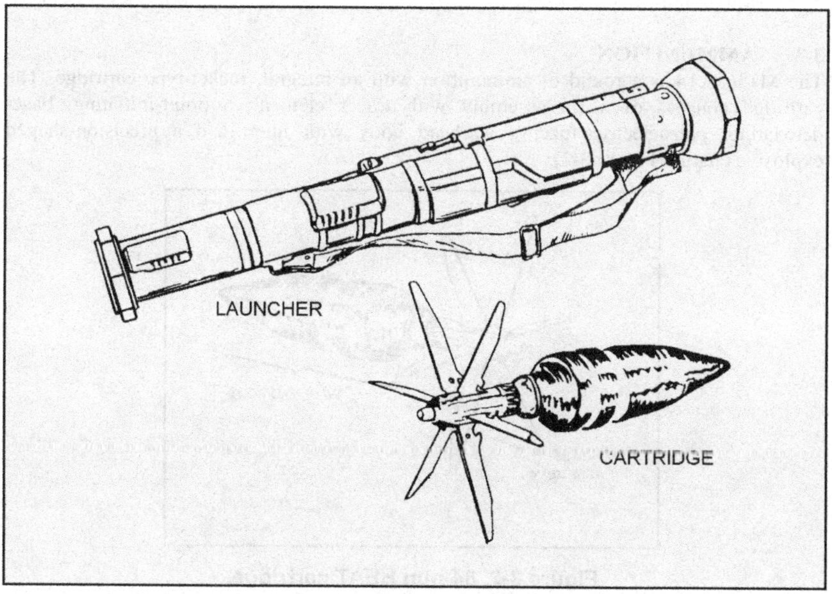


**Figure 3-1. Launcher and HEAT cartridge.**

## 3-2. TECHNICAL DATA

The following data apply to the M136 AT4:

a. **Launcher.**

Length ........ 1,020 mm (40 inches)
Weight (Complete System) ........ 6.7 kg (14.8 pounds)
Rear Sight ........ Range indicator, graduated in 50-meter increments

b. **Rocket.**

Caliber ........ 84 mm
Muzzle Velocity ........ 290 mps (950 fps)
Length ........ 460 mm (18 inches)
Weight ........ 1.8 kg (4 pounds)
Minimum Range
Training ........ 30 meters (100 feet)
Combat ........ 10 meters (33 feet)
Arming ........ 10 meters (33 feet)
Maximum Range ........ 2,100 meters (6,890 feet)
Maximum Effective Range ........ 300 meters (985 feet)

## 3-3. AMMUNITION

The M136 AT4 is a round of ammunition with an integral, rocket-type cartridge. The cartridge consists of a fin assembly with tracer element; a point-initiating, base-detonating, piezoelectric fuze; a warhead body with liner; and a precision-shaped explosive charge (Figure 3-2).

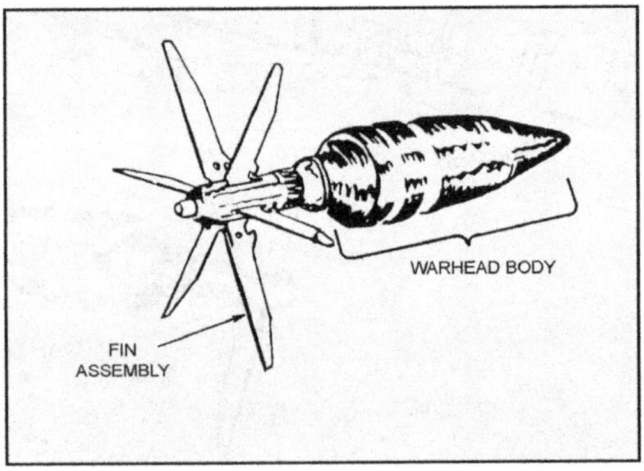


**Figure 3-2. 84-mm HEAT cartridge.**

a. **Description.** The M136 AT4's warhead has excellent penetration ability and lethal after-armor effects. The extremely destructive, 440 gram shaped-charge explosive penetrates more than 14 inches (35.6 cm) of armor. Warhead effects are shown in Figure 3-3.

(1) ***Impact***. The nose cone crushes; the impact sensor activates the fuze.

(2) ***Ignition***. The piezoelectric fuze element activates the electric detonator. The booster detonates, initiating the main charge.

(3) ***Penetration***. The main charge fires and forces the warhead body liner into a **directional gas jet that penetrates armor plate.**

**(4)** ***After-Armor Effects (Spalling)***: The projectile fragments and incendiary effects produce blinding light and destroy the interior of the target.

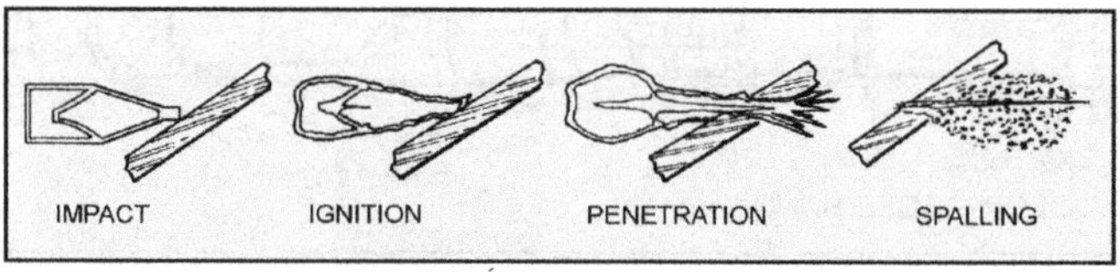


**Figure 3-3. Effects of M136 AT4 warhead.**

b. **Packaging**. Five M136 AT4s, each wrapped in a plastic barrier bag, are packed together in a wooden container. The containers are too heavy to stack more than four deep on the pallets (Figure 3-4).

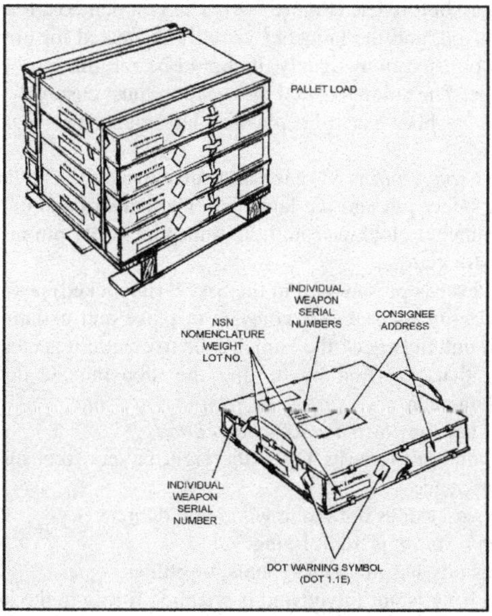


**Figure 3-4. Ammunition packaging and markings.**

c. **Color-Coding**. M136 AT4 launchers are marked with color-coded bands (Figure 3-5). A black with yellow band indicates an HE antiarmor round (early models had a solid black band). A gold or yellow band indicates a field handling trainer; no band indicates an M287 9-mm tracer bullet trainer (Appendix B).

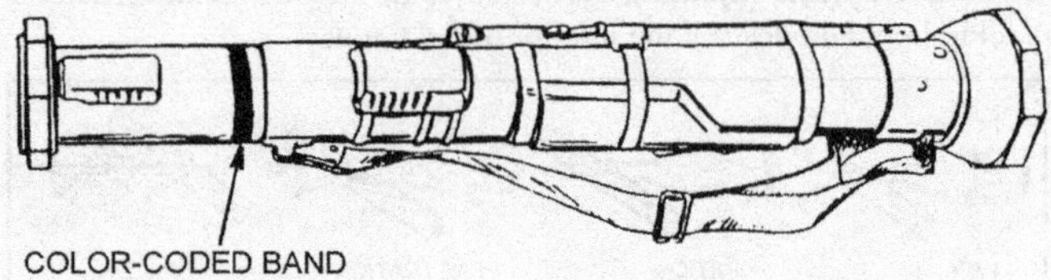


**Figure 3-5. Location of color-coded band.**

## 3-4. INSPECTION

Since the M136 AT4 is issued as a round of ammunition rather than as a weapon, the launcher is completely sealed. However, its overall condition should be inspected at the time of issue and again before use (Figure 3-6). The wooden container should be opened, the plastic bags removed, and the launcher visually inspected for obvious damage. If the M136 AT4 is not to be used immediately, it should be returned to its plastic bag and the bag resealed with tape. The soldier issued the weapon must ensure—

- The *rear seal*, a brown acrylic plastic plate inside the *venturi*, is in place and undamaged.
- The *transport safety pin* is in place and fully inserted. The lanyard is attached to the transport safety pin and the launcher. The lanyard should already be wrapped around the launcher clockwise and the transport safety pin inserted in the retainer hole counterclockwise.
- The *cocking lever* is present and in the SAFE (uncocked) position.
- The plastic *fire-through muzzle cover* is in place and undamaged. If it is torn or broken, cut it out and check the launch tube to ensure it is clear of foreign objects. Remove any that you find by turning the tube muzzle downward and gently shaking the launcher.
- The launcher has the correct *color-coded band*.
- The *sights* function properly. Open the sight covers to ensure the sights pop up and are undamaged.
- The *forward safety* does not move when you depress it.
- The *red trigger button* is not missing.
- The launcher body has no cracks, dents, or bulges.
- The *carrying sling* is not frayed and is attached firmly to the launch tube.
- The *shoulder stop* is not broken or damaged, and it unsnaps and folds down.

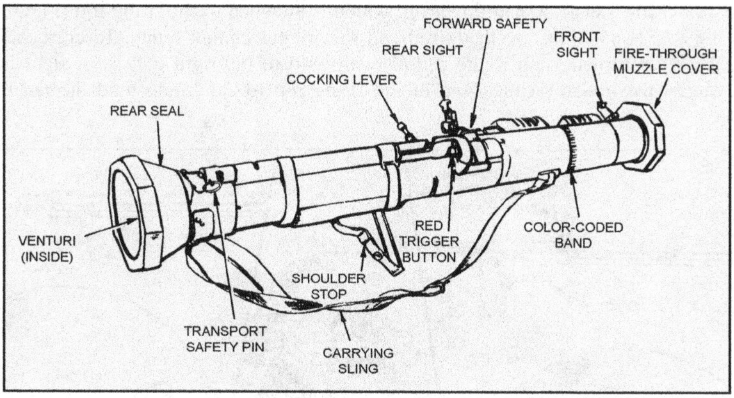


**Figure 3-6. Inspection.**

## 3-5. FIRING MECHANISM, SAFETIES, AND WEAPON FUNCTION

The function of the M136 AT4 must be discussed along with its firing mechanism and safety features. The firing mechanism is mechanical and consists of a red trigger button, an enclosed firing rod and spring, and three safety devices (Figure 3-7). The first safety device is the *transport safety* pin. The red trigger button is located between the other two, the *cocking lever* and the *forward safety*. The weapon cannot be fired until all three safeties have been disengaged.

a. **Transport Safety Pin**. This pin blocks the firing pin from striking the cartridge percussion cap. To disengage this pin, pull it outward, then release it.

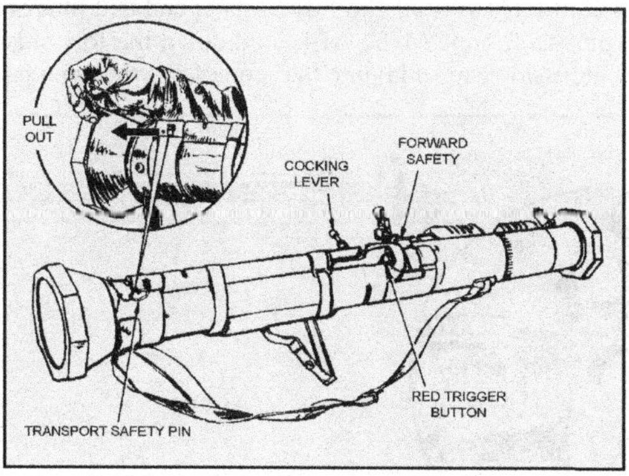


**Figure 3-7. Firing mechanism and transport safety pin.**

b. **Cocking Lever**. When this lever, which is attached to the firing rod (Figure 3-8), is in the SAFE position, the firing rod and the trigger cannot touch. To cock the AT4, push the lever forward and rotate it downward and to the right with your right thumb. This causes the hooks on the front of the firing rod to catch and hold the red trigger button.

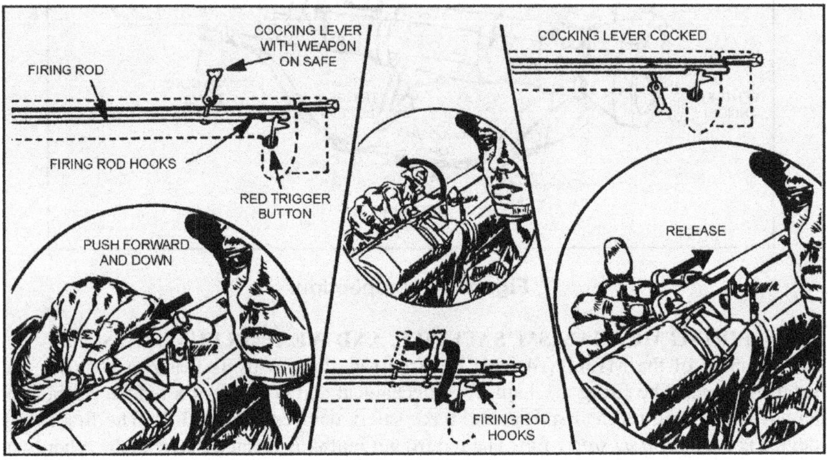


**Figure 3-8. Cocking lever.**

c. **Forward Safety**. This safety is on the front end of the firing mechanism (Figure 3-9) and is connected to a steel rod with a bent end that blocks the firing rod from striking the firing pin. To fire the M136 AT4, hold down the forward safety so the firing rod can strike the percussion cap and ignite the propellant when you push the trigger.

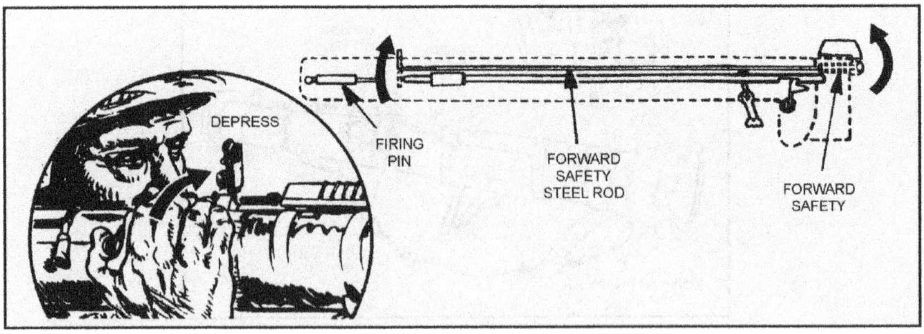


**Figure 3-9. Forward safety.**

## 3-6. SIGHTS

The fact that the AT4's front and rear sights resemble those of the M16-series rifle makes using the AT4 easier (Figure 3-10).

a. **Front Sight**. The front sight has a sight blade with a center post and left and right lead posts. A semicircular white line helps you obtain the proper sight picture. To open the front sight cover, press down on it and slide it backward until the sight pops up.

b. **Rear Sight**. The rear sight has a sight blade, range adjustment knob, range scale, 2-mm peephole for normal daylight visibility conditions, and 7-mm peephole for limited visibility conditions. To open the rear sight cover, press down on it and slide it forward until the sight pops up.

c. **Nightsight**. The AT4 can be fitted with the AN/PAQ-4C, AN/PEQ-2, or the AN/PAS-13, when used with the nightsight mounting bracket (NSN 5340-01-391-3004).

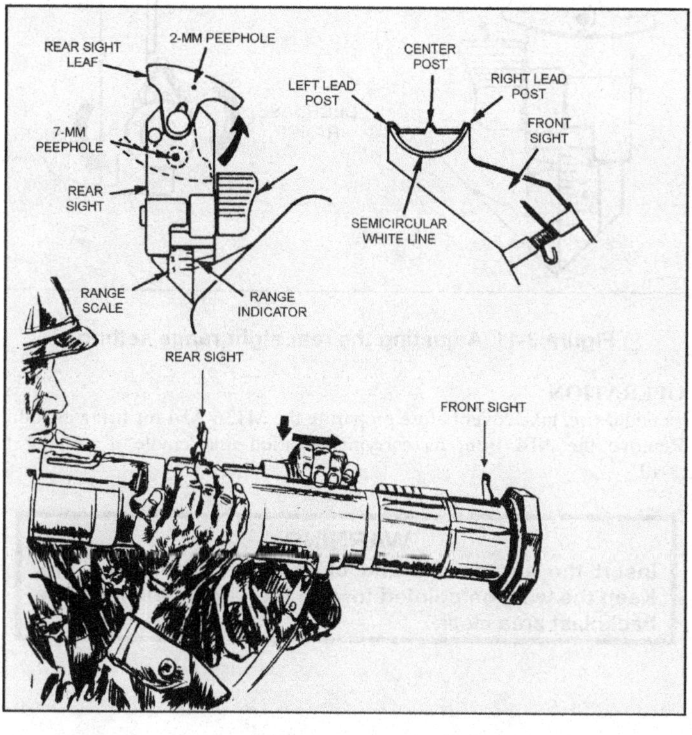


**Figure 3-10. Sights.**

(1) The leaf blade that covers the 7-mm peephole has its own tiny 2-mm peephole. To uncover the 7-mm peephole, pull the bottom of the leaf blade out slightly and rotate it right and up. To cover the 7-mm peephole, rotate it back down and ensure the leaf blade

is seated. The range indicator scale is indexed from 100 to 500 meters in 50-meter increments.

(2) To increase the range setting beyond 200 meters, turn the range adjustment knob clockwise, or vice versa (Figure 3-11). You must remember to reset the range to 200 meters when you close the rear sight. Otherwise, closing the sight cover will break off the rear sight.

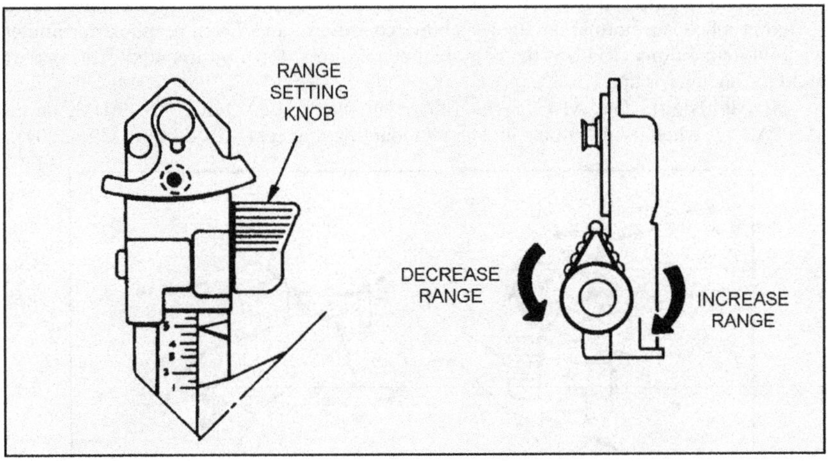


**Figure 3-11. Adjusting the rear sight range setting.**

## 3-7. OPERATION

If you are under fire, take cover before preparing the M136 AT4 for firing as follows:

a. Remove the AT4 from its carrying position and cradle it in your left arm (Figure 3-12).

> **WARNING**
> **Insert the approved brand of earplugs before you fire. Keep the weapon pointed toward the target, and keep the backblast area clear.**

Figure 3-12. Cradle position.

b. With your right hand, pull and release the transport safety pin (Figure 3-13). This pin is important—you must reinsert it if you do not fire the launcher. Therefore, unless it is attached to the launcher with a lanyard, you must keep it in a safe place.

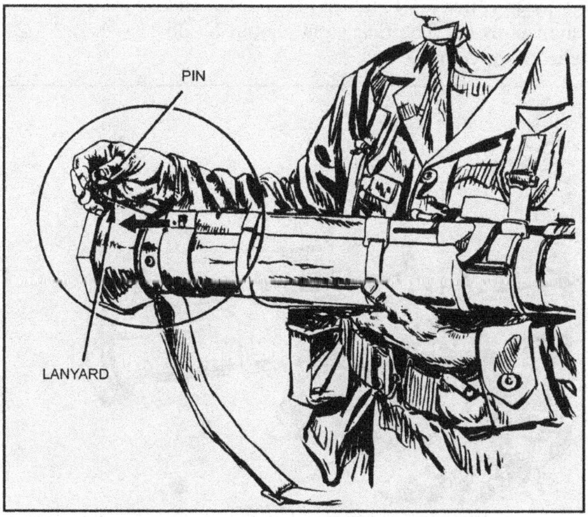


Figure 3-13. Removing the transport safety pin.

c. Unsnap, unfold, and hold the shoulder stop with your right hand (Figure 3-14).

d. Place the launcher on your right shoulder and stabilize it by grasping the sling near the launcher's muzzle with your left hand.

**Figure 3-14. Unsnapping the shoulder stop.**

e. With the AT4 on your right shoulder, stabilize it with your left hand and open the sights with your right hand. Press down and pull backward on the front sight cover until the front sight pops up (Figure 3-15), then press down and forward on the rear sight cover until the rear sight pops up. The rear sight should be no less than 2 1/2 inches and no more than 3 inches from your eyes.

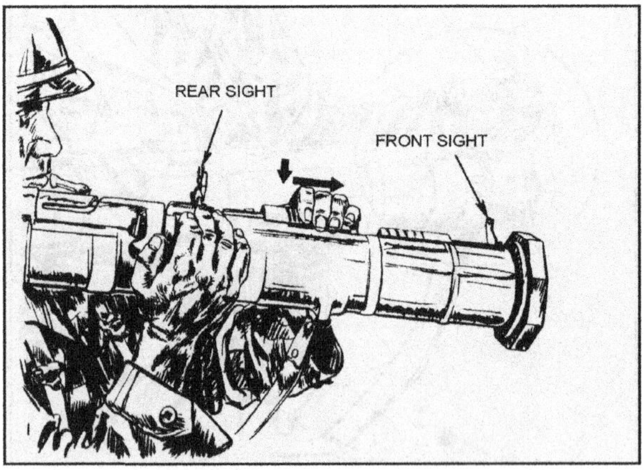


**Figure 3-15. Opening and adjusting the sights.**

f. Set the rear sight for the correct range to the target.

g. Check the backblast area before you cock the launcher. Then, with your right hand, unfold the cocking lever (Figure 3-16). Place your thumb under it and, with the support of your fingers in front of the firing mechanism, push it forward, rotate it downward and to the right, and let it slide backward.

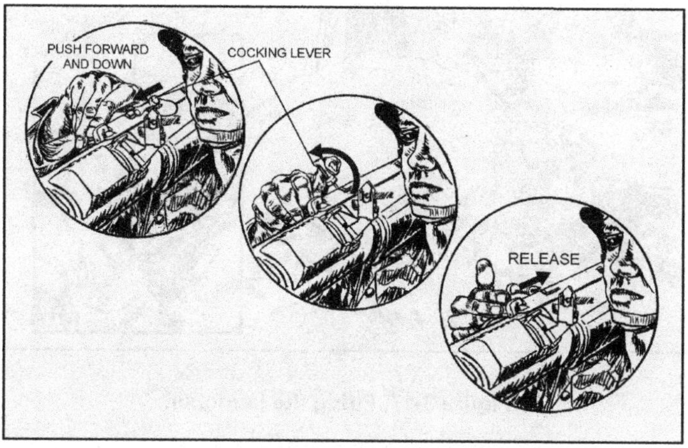


**Figure 3-16. Cocking the launcher.**

h. Pull back on the sling with your left hand to seat the shoulder stop firmly against your shoulder. To avoid a misfire, use the index and middle fingers on your right hand to hold the forward safety down and to the left while you fire (Figure 3-17, page 3-12).

## 3-8. MISFIRE PROCEDURES

A misfire is a complete failure to fire caused by a procedural or mechanical failure. Which misfire procedures should be used depends on whether the firer is in a combat or training environment.

a. **Causes.** A misfire is usually caused by one of the following factors:

The forward safety is not depressed far enough to disengage the safety.

The firing mechanism is faulty.

The propelling charge explosive train is faulty.

**WARNING**
**Keep your weapon pointed toward the target.**

**Figure 3-17. Firing the launcher.**

b. **Combat Environment**. If a misfire occurs in combat—

(1) Release the forward safety.

(2) Remove your right hand from the firing mechanism and cock the weapon again.

(3) Try to fire again. If the launcher still does not fire, maintain the same firing position and return the cocking lever to the SAFE (uncocked) position.

(4) Move the launcher from your shoulder, keeping the launcher pointed toward the enemy. Reinsert the transport safety pin.

(5) In combat, break off the sights to identify the misfired launcher. In training, however, you would not want to damage the field handling trainer (FHT), so identify the misfired launcher simply by leaving the sights up.

(6) Place the launcher on the ground, pointed toward the enemy, and use another launcher. As soon as you can, dispose of the misfired launcher IAW unit SOP.

c. **Training Environment**. If a misfire occurs on a live-fire training range, the soldier responds as follows (the trainer later disposes of the launcher IAW local SOP):

(1) Shouts "Misfire" as soon as the launcher fails to fire while maintaining the original sight picture.

(2) Releases the forward safety.

(3) Recocks the launcher: Immediately removes his right hand from the firing mechanism and pushes the cocking lever forward with the heel of his right hand until the lever locks with a loud clicking noise.

**NOTE:** Because performing immediate action takes so little time, you need not recheck the backblast area.

(4) Press the forward safety all the way down and try to fire again. If the launcher still fails to fire, shout misfire, release the forward safety, and move the cocking lever to the SAFE (uncocked) position. Move the launcher from your shoulder, keeping the weapon pointed toward the target and cradled in your left arm.

(5) Reinsert the transport safety pin, wait two minutes, then carefully lay the launcher on the ground, muzzle toward the target.

**NOTE:** Notify the local ammunition supply and issue point of any unusual occurrence, regardless of whether the weapon fires or not. Examples include excessive overpressure, recoil, or heat on your face after you have fired the weapon (caused by the propellant burning after the round leaves the muzzle).

## 3-9. RESTORATION TO CARRYING CONFIGURATION

If the launcher is prepared to fire, but then is not fired, it must be taken out of operation as follows:

a. Release the forward safety.

b. Push forward and to the left on the cocking lever, and let it spring back into the SAFE (uncocked) position.

c. Move the launcher from your shoulder, ensuring the muzzle is pointed in the direction of fire.

d. With the launcher cradled in your left arm, replace the transport safety pin until it is fully seated in the retainer hole.

e. To avoid breaking off the rear sight, remember to reset the range indicator to the 200-meter setting before closing the rear sight cover.

f. Lay down the sights and close their covers.

g. Snap the shoulder stop into the closed position.

h. Sling the launcher over your right shoulder and move to another location.

CHAPTER 4
# MARKSMANSHIP FUNDAMENTALS

*Many factors contribute to light antiarmor weapon marksmanship. Soldiers who combine these factors well, and continue to practice doing so, can retain their skills. The factors are grouped into four basic areas known as marksmanship fundamentals: steady hold, aiming procedures, breath control, and trigger manipulation. Instructions are given for right-handed firers, but the M72-series LAW can be fired from either shoulder by simply reversing the instructions.*

## 4-1. STEADY HOLD

Maintaining a steady hold involves holding the launcher as steady as possible while sighting and firing. To maintain the proper sight picture and sight alignment until he fires, the firer must hold the launcher in a tight, comfortable position so that it becomes a natural extension of his body (Figure 4-1). With both weapons, keep your elbows close to your body to help balance the weapon and prevent you from jerking or flinching when you fire. In the case of the AT4, this reduces recoil.

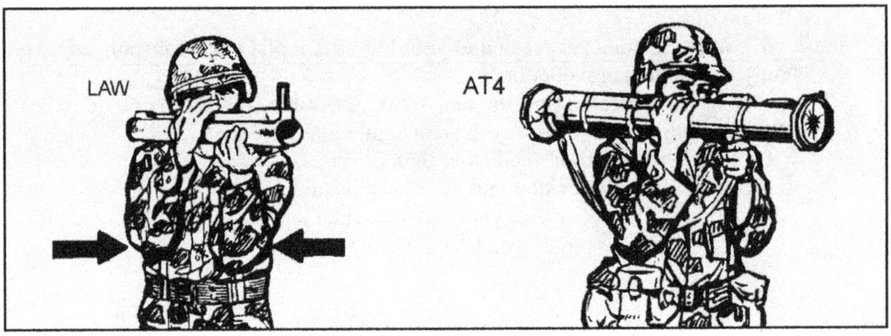


**Figure 4-1. Steady hold position.**

a. **M72-Series LAW**. Place your left hand, palm facing upward, under the launcher near the muzzle and grasp the launcher. Firmly pull the rear cover into your right shoulder pocket.

b. **M136 AT4**. With your left hand, grasp the carrying sling where it attaches to the launcher near the muzzle. With your right hand on the trigger mechanism, pull the shoulder stop into your right shoulder pocket.

## 4-2. AIMING PROCEDURES

Aiming procedures include placing the eye correctly, obtaining a sight picture, and aligning the sight. Combining these procedures is critical to correctly aiming light antiarmor weapons.

a. **Eye Placement**. Before sighting the weapon, estimate the range (Chapter 6 discusses range estimation). For the M72-series LAW, place your firing eye as close to the rear sight as is comfortable. However, for the M136 AT4, place your firing eye between 2 1/2 to 3 inches (no nearer than 2 1/2 inches) from the rear sight. This distance is necessary to prevent possible injury from the weapon's recoil and for correct sight alignment with the AT4 (Figure 4-2).

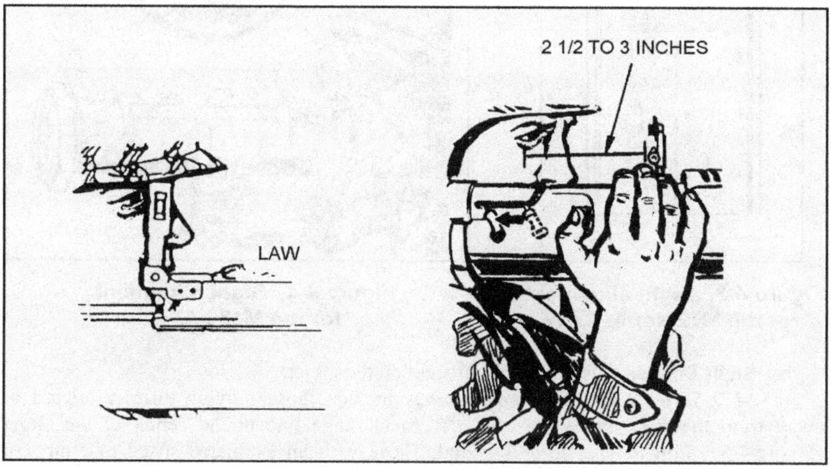


**Figure 4-2. Eye placement.**

> **WARNING**
> **When firing the M136 AT4, do not place your eye within 2 1/2 inches of the rear sight. The AT4s recoil could cause the rear sight to injure your firing eye.**

b. **Sight Alignment.** Align the sights correctly with the target. To do this for the M72-series LAW, position the rear sight so that your eye is near and in line with the peephole in the rear sight. Look through the peephole at the front sight reticle and place the range line that corresponds to the target's range on the target (Figure 4-3). To do this for the AT4, position the rear sight so that the white semicircle of the front sight is a hazy line around the bottom half of the rear sight opening. Position the front sight posts on the target (Figure 4-4). Align the sight by moving your head forward or backward.

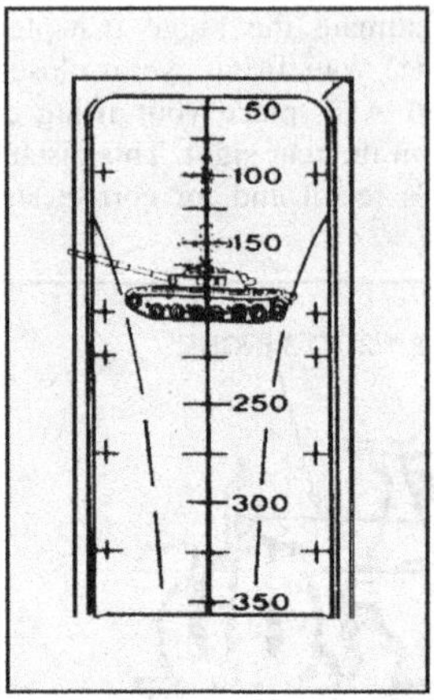


Figure 4-3. Sight alignment for the M72-series LAW.

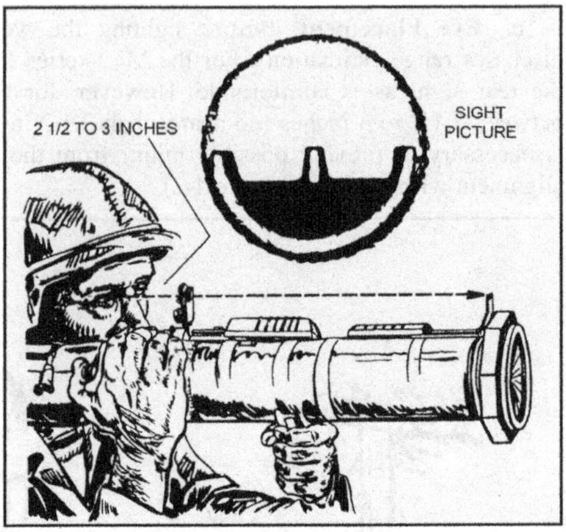


Figure 4-4. Sight alignment for the M136 AT4.

c. **Sight Picture**. Position the front sight on the target.

(1) ***M72-Series LAW***. *Stationary targets* include those moving directly toward or away from the firer. Place the correct vertical range line in the center of the target (Figure 4-5). *Slow-moving targets* include those with an estimated speed of 5 mph or slower, or those moving in an oblique direction. Place either the left or right lead cross mark on the vehicle's center of mass (Figure 4-6). *Fast-moving targets* include those with an estimated speed of more than 5 mph. Place either the left or right lead cross mark on the leading edge of the vehicle (Figure 4-7).

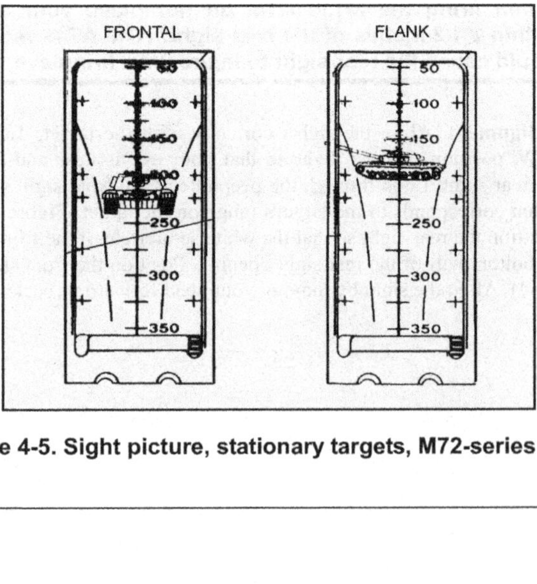


Figure 4-5. Sight picture, stationary targets, M72-series LAW.

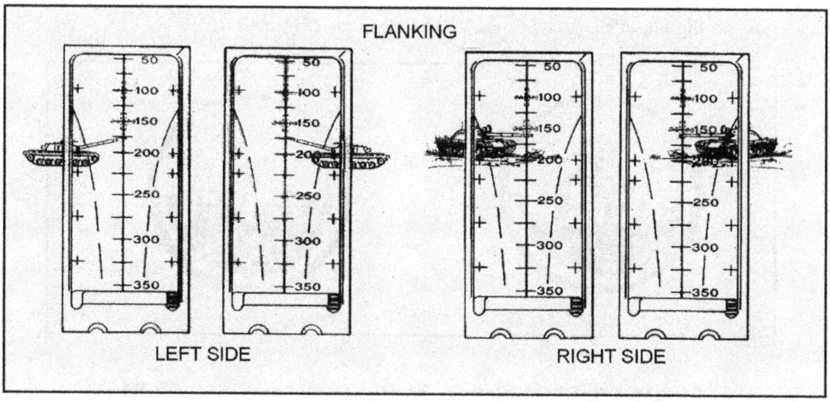


**Figure 4-6. Sight picture, slow-moving targets, M72-series LAW.**

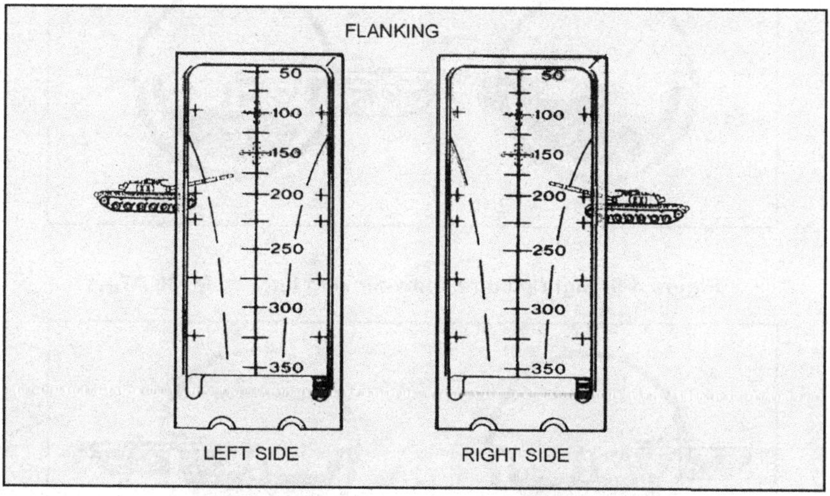


**Figure 4-7. Sight picture, fast-moving targets, M72-series LAW.**

(2) ***M136 AT4***. *Stationary targets* include those moving directly toward or away from the firer. Adjust the rear sight for the correct range and place the center sight post in the center of the target (Figure 4-8). *Slow-moving vehicles* are those with an estimated speed of 10 mph or less, or those moving in an oblique direction. Place the center sight post on the front or leading edge of the vehicle (Figure 4-9). *Fast-moving vehicles* are those estimated to be moving faster than 10 mph. Place either the left or right lead post on the

center of the target. For example, if the target is moving from left to right, place the left lead post on the target's center of mass, and vice versa (Figure 4-10).

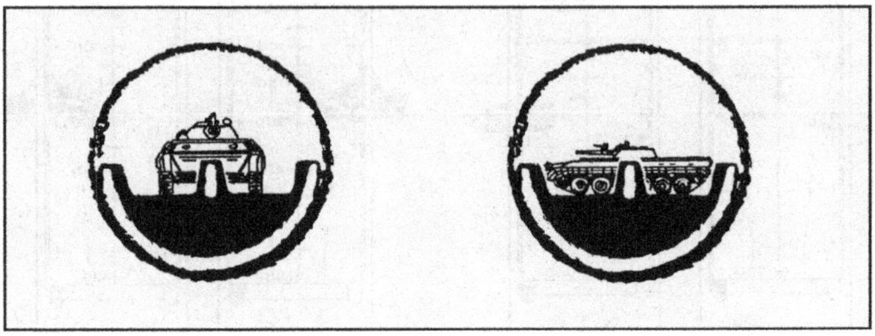

**Figure 4-8. Sight picture, stationary targets, M136 AT4.**

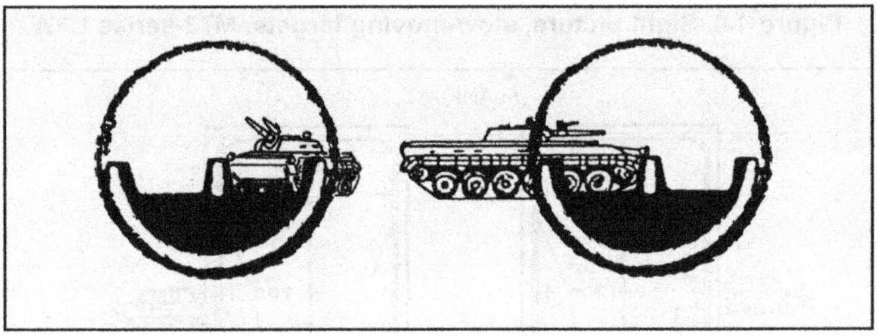

**Figure 4-9. Sight picture, slow-moving targets, M136 AT4.**

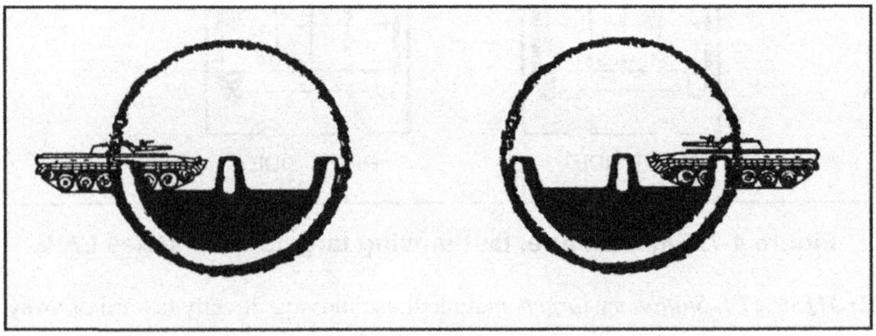

**Figure 4-10. Sight picture, fast-moving targets, M136 AT4.**

## 4-3. BREATH CONTROL

Breath control is as important when firing a light antiarmor weapon as it is when firing an individual weapon. Breathing while firing can cause a miss. To control breathing, the firer breathes deeply a couple of times, takes one last deep breath, exhales partly, holds his breath, sights, and fires.

## 4-4. TRIGGER MANIPULATION

Light antiarmor weapons have different types of triggers (Figure 4-11).

a. **LAW**. To fire the LAW, the firer must apply firm and steady downward pressure to the trigger with the fingers of his firing hand.

b. **AT4**. To fire the AT4, the firer must apply firm and steady forward pressure to the trigger with the thumb of his firing hand. Soldiers can practice trigger manipulation and control techniques on an expended launcher or FHT.

**Figure 4-11. Trigger manipulation.**

## 4-5. INTEGRATED ACT OF SHOOTING

Correct sight alignment is critical. Sight alignment errors increase as the range to the target increases. Therefore, maintaining the correct relationship between the rear and front sights is as important as placing the aiming point. The steps for doing this should become automatic. No matter how quickly they are done, these steps are always distinct, because the human eye can only focus at one distance and on one point at a time. The firer focuses on the front sight to obtain correct sight alignment, then places the aiming point to complete the sight picture. He shifts or adjusts the position of the launcher as necessary. The whole time he is pressing the trigger, he maintains the sight picture.

## CHAPTER 5
# FIRING POSITIONS

*This chapter explains the basic firing positions used with light antiarmor weapons. Instructions for each are given for right-handed firers, but the M72-series LAW can be fired from either shoulder by simply reversing the instructions. Though each weapon can be fired from all four of the basic firing positions, individual physique determines exact body and hand positions. Firing from a supported position naturally increases accuracy, which improves the odds for a first-round hit or kill. Basic safety considerations are the same for all light antiarmor weapons, but additional considerations for each firing position are provided here.*

### 5-1. STANDING POSITION

Two standing positions are used: a basic standing position and one modified for the infantry fighting position.

a. **Basic Standing Position**. Raise the launcher slightly higher than shoulder level. Execute a left face, rotate your shoulder under the launcher, and spread your feet a comfortable distance apart. Move your left foot 15 to 24 inches forward, keeping your hips level and your weight balanced on both feet. To obtain a firm, stable position, tuck both elbows tightly into your body. To track a moving target, turn your body at the waist—not with your legs. This enables you to track the target smoothly. Unless you are behind a protective barrier such as a wall, the standing position exposes you more than any other position to enemy observation and possible suppression. Differences between weapons with respect to the standing position are as follows (Figure 5-1, page 5-2):

**WARNING**
**Always keep the launcher pointed in the direction of fire.**

(1) ***M72-Series LAW***. Place your nonfiring hand about 4 inches from the front of the muzzle, with your firing hand on the rear cover. After placing the weapon on your shoulder, release the rear cover and place your firing hand on the trigger. Cup the launcher in the palm of your nonfiring hand. Position your firing eye as close to the rear sight as is comfortable.

(2) ***M136 AT4***. Grasp the sling near the launcher with your left hand and the shoulder stop with your right hand. Raise the launcher above shoulder level. After placing the launcher on your shoulder, release the shoulder stop and place your right hand on the trigger. Place your firing eye 2 1/2 to 3 inches from the rear sight.

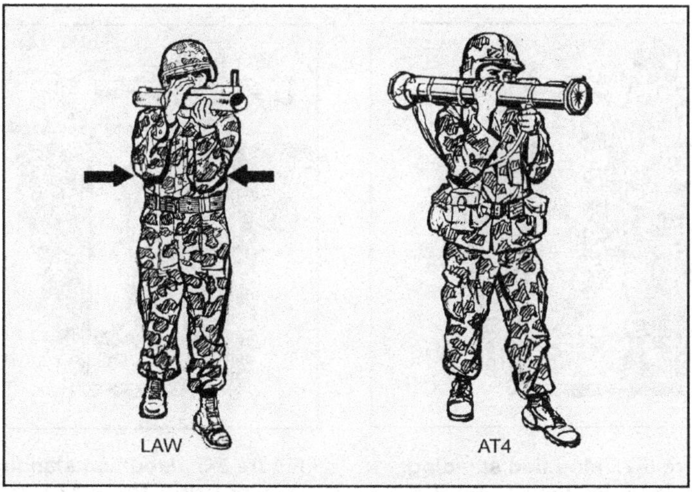


**Figure 5-1. Basic standing position.**

b. **Modified Standing Position**. Use this position when you occupy an infantry fighting position. Assume the basic standing position, but instead of stepping forward, lean against the back wall of the fighting position. Ensure that the venturi or rear of the weapon extends beyond the rear of the fighting position. Figure 5-2 shows the modified standing position for the M72-series LAW; Figure 5-3 shows the modified standing position for the M136 AT4. Ensure that NONE of the following are in your backblast area:

- Other soldiers.
- Other fighting positions.
- Equipment.
- Any part of your own fighting position.
- Obstructions within 5 meters.

NOTE: Leaders must ensure that light antiarmor weapons are positioned so that the backblast misses other fighting positions.

**DANGER**
**NEVER FIRE FROM WITHIN A COMPLETELY ENCLOSED, UNVENTILATED BUNKER OR FIGHTING POSITION.**

Figure 5-2. Modified standing, position, M72-series LAW.

Figure 5-3. Modified standing position, M136 AT4.

## 5-2. KNEELING POSITION

The basic kneeling position is the best position for tracking moving targets. The modified kneeling position is best for engaging stationary targets, since it is a supported position. However, either can be used for stationary or moving targets.

a. **Basic Kneeling Position.** Kneel from the basic standing position onto your right knee, keeping your left thigh parallel to the ground. Rotate your lower right leg 90 degrees to the left. (This removes your right foot from exposure to the backblast.) Keep your right thigh and back straight and perpendicular to the ground. Point your left foot in the direction of fire and tuck your elbows in to your sides. Though this is not a supported position, it should be a firm, stable one. Figure 5-4 shows the basic kneeling positions for the LAW and AT4.

b. **Modified Kneeling Position.** From the basic kneeling position, sit back on your right heel. Place the back of your upper left arm on your left knee, making sure you do not have bone-to-bone contact between your left elbow and left knee. Keep your right elbow tucked in close to your right side. Use any protective barriers available. Figure 5-5 shows the modified kneeling positions for the LAW and AT4.

**Figure 5-4. Basic kneeling positions.**

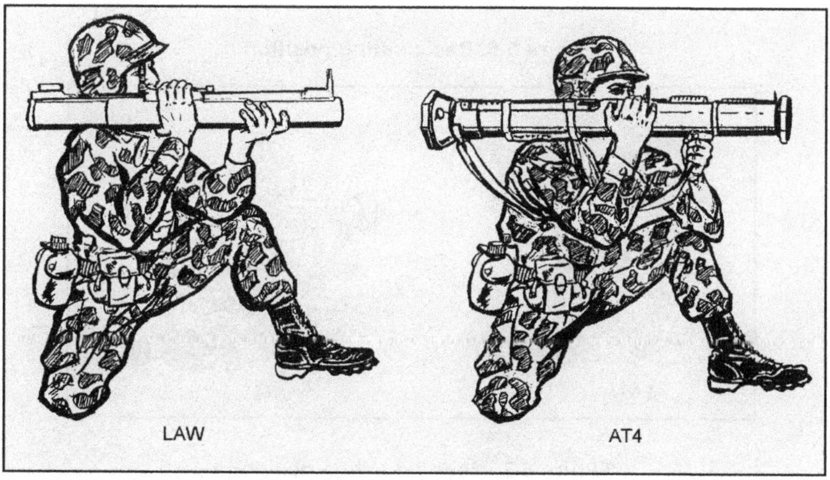


**Figure 5-5. Modified kneeling positions.**

## 5-3. SITTING POSITION

The sitting position is the most stable firing position. In this position, the arms are placed on the legs for support. Depending on his physique, the firer can use either of two versions of the sitting position. Either is suitable for engaging stationary targets.

a. **Basic Sitting Position.** Sit on your buttocks while facing the target, and spread your feet a comfortable distance apart. Lean forward and place the backs of your upper arms on your knees, avoiding bone-to-bone contact. Figure 5-6 shows the basic sitting positions for the LAW and AT4.

b. **Modified Sitting Position.** From the basic sitting position, cross your ankles for added support. Raise or lower your knees to adjust for elevation on the target. Figure 5-7 shows the modified sitting positions for the LAW and AT4.

**Figure 5-6. Basic sitting positions.**

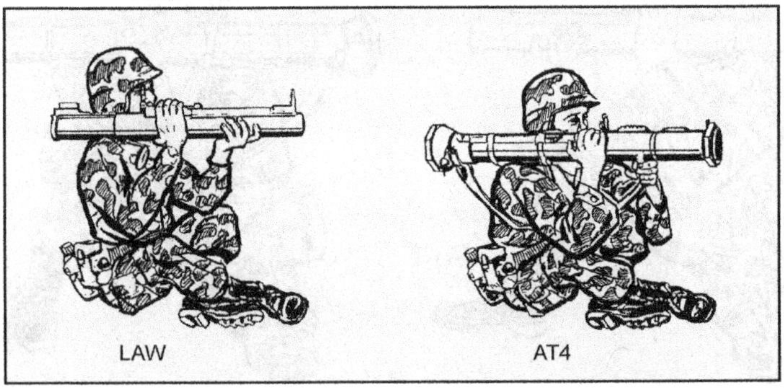


**Figure 5-7. Modified sitting positions.**

## 5-4. PRONE POSITION

The prone position is the most dangerous position due to its proximity to the ground. Ideally, the ground should slope downward from the rear of the launcher. This slope reduces the effects of the backblast.

a. Lie on your stomach with your body at a 90-degree angle to direction of fire, and with your body and legs to the left of the direction of fire.

b. *Ensure that neither the body nor the legs are in the backblast area.*

c. Unlike other firing positions, this one prevents you from placing the launcher on your right shoulder. Instead, you must hold the launcher in place against your upper right arm. For stability, apply extra pressure on the firing mechanism with your right hand. The prone position is the *least* stable of all firing positions. You must practice it often to become confident using it. Figure 5-8 shows the prone positions for the LAW and AT4.

**DANGER**

**FAILURE TO MAINTAIN A 90-DEGREE ANGLE FROM THE DIRECTION OF FIRE COULD CAUSE INJURY OR DEATH TO THE FIRER.**

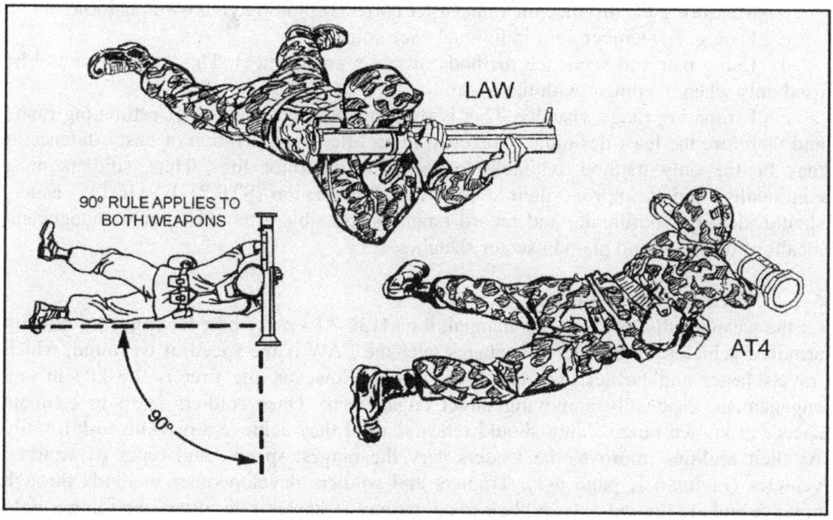


**Figure 5-8. Prone positions.**

CHAPTER 6

# COMBAT TECHNIQUES

*This chapter discusses employment techniques for light antiarmor weapons, all of which require at least basic gunnery skills. Techniques that require advanced skills are identified as such.*

## 6-1. RANGE ESTIMATION

A firer who can accurately estimate the range to the target has a better chance of hitting it, regardless of the weapon used. Common methods of estimating range are listed below from the most to the least accurate. The tactical situation determines the method to be used:

a. Using range finders.

b. Measuring the distance on a map after correctly plotting your own position.

c. Pacing. Remember your individual pace count.

d. Using pair and sequence methods of target engagement. This method should be used only when in contact with the enemy.

e. Estimating range visually. This is the least accurate method of estimating range and therefore the least desirable. However, in an offensive operation or hasty defense, it may be the only method available to the light antiarmor firer. Thus, soldiers must continually train to improve their skill at visual estimation (STP 21-1-SMCT). Leaders should identify, coordinate, and record ranges to possible armored vehicle engagement locations on squad and platoon sector sketches.

## 6-2. SPEED ESTIMATION

Of the weapons discussed in this manual, the M136 AT4 is the best for engaging moving armored vehicles. One of its advantages over the LAW is the speed of its round, which travels faster and farther than the LAW round. However, the firer is the key in any engagement, especially a moving target engagement. Once soldiers learn to estimate speeds at known ranges, they should rehearse until they achieve a high hit-to-kill ratio. As their abilities improve, the leaders vary the ranges, speeds, and types of armored vehicles (Figure 6-1, page 6-2). Trainers and soldiers develop other methods through practice and are limited only by their imaginations. (Chapter 4 discusses obtaining a sight picture in detail.) Estimate how far the vehicle travels in 1 second:

a. Start when the front end of the vehicle passes the object.

b. Count, "One thousand and one" (takes about one second).

c. If *more* than half of the vehicle passes the object, estimate it as a *fast-moving* vehicle (10 mph or faster). If *less* than half of the vehicle passes the object, estimate it as a *slow-moving vehicle* (less than 10 mph).

Figure 6-1. Speed estimation.

## 6-3. ARMORED VEHICLE WEAKNESSES

Armored vehicles usually have their heaviest armor in front, because they are designed mainly for offensive operations against other armored vehicles (Figure 6-2). All vehicles are vulnerable to repeated hits on their flanks and rear, though the flank offers the largest possible target. Firers should always aim center of mass to increase the probability of a hit. The older the vehicle model, the less protection it has against antiarmor weapons. Consequently, newer versions may use bolt-on (applique) armor to improve their survivability. Some vehicles are equipped with reactive armor, which consists of metal plates and plastic explosives. Reactive armor usually covers the forward-facing portions and sides of the vehicle and can defeat shaped-charge weapons such as the LAW and AT4. When reactive armor detonates, it disperses metal fragments to 200 meters. The M72-series LAW and the M136 AT4 cause only a small entry hole in an armored vehicle target, though some fragmentation or spall may occur.

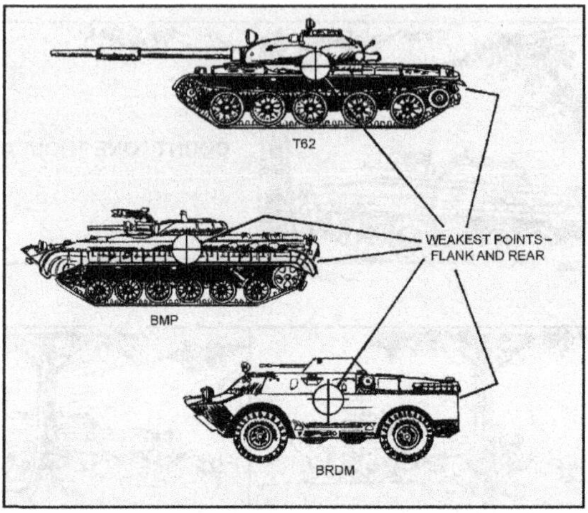


**Figure 6-2. Armored vehicle weak points.**

a. Natural or man-made obstacles can be used to force the armored vehicle to slow, stop, or change direction. This pause enables the firer to achieve a first-round hit. If he does not achieve a catastrophic kill on the first round, he or another firer must be ready to engage the target vehicle immediately with another round.

b. An armored vehicle without close protection (dismounted infantry) in woods, MOUT, or other restrictive terrain is vulnerable to close attack. This type of attack is most likely to originate from well-armed infantry-type teams organized into armor-killer teams. (Noninfantry units may also be required to perform this mission.) Skilled firers from these teams should engage the suspension or engine compartment of vehicles that have applique or reactive armor. When an armored vehicle is buttoned up—all hatches are closed and personnel are inside the vehicle—the crew cannot see well enough to protect itself from close attacks or attacks from the flanks or rear. The personnel inside cannot see anything within 10 meters of the vehicle, and they cannot shoot at anything (using their main guns) within 20 meters. The white area in Figure 6-3, page 6-4 shows the most favorable direction of attack when the turret is facing to the front; the gray area shows the vehicle's principal direction of fire and observation when the turret is facing to the front.

c. Armored vehicle kills are classified according to the level of damage achieved (Table 6-1).

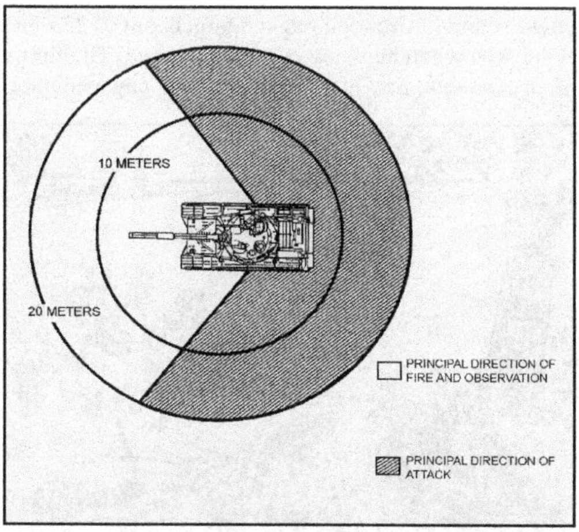


**Figure 6-3. Limited visibility of armored vehicles.**

| TYPE OF KILL | PART OF VEHICLE DAMAGED OR DESTROYED | CAPABILITY AFTER KILL |
|---|---|---|
| Mobility Kill | Suspension (track, wheels, or road wheels) or power train (engine or transmission) has been damaged. | Vehicle cannot move, but it can still return fire. |
| Firepower Kill | Main armament has been disabled. | Vehicle can still move, so it can get away. |
| Catastrophic Kill | Ammunition or fuel storage section has been hit by more than one round. | Vehicle completely destroyed. |

**Table 6-1. Armored vehicle kills.**

## 6-4. METHODS OF ENGAGEMENT

The four engagement methods include single, sequence, pair, and volley firing. The leader evaluates the situation on the ground to determine which of these methods to use. Regardless of whether they are used singly or in combination, communications are needed as well. The methods of engagement are rehearsed IAW unit SOP.

a. **Single Firing**. A single soldier with one light antiarmor weapon may engage an armored vehicle, but this is not the preferred method of engagement. Several light antiarmor weapons are required to kill an armored vehicle. A single firer firing one round must hit a vital part of the target to damage it at all (Figure 6-4).

(1) ***Range Not Known***. A single firer should engage only targets within 200 meters when he does not know the actual range. The probability that he hits a target beyond 200 meters with a single round is small.

(2) ***Range Known***. A single firer can engage targets out to 225 meters with the LAW or 300 meters with the AT4 when he knows the actual range. The firer should do this only when he has a flank or rear shot, or when he has no other engagement option.

**Figure 6-4. Single firing.**

b. **Sequence Firing**. A single firer, equipped with two or more light antiarmor weapons prepared for firing, engages the target. After engaging with the first round and observing the impact, the firer adjusts his point of aim, engages with another round, and so on until he destroys the target or runs out of rounds (Figure 6-5).

Figure 6-5. Sequence firing.

c. **Pair Firing**. Two or more firers, equipped with two or more light antiarmor weapons prepared for firing, engage a single target. Before firing, the first firer informs the others of the estimated speed and distance to the target. If the impact of his round proves his estimate to be correct, the other firers engage the target until it is destroyed. If the impact of the round proves his estimate to be incorrect, the second firer informs the others of his own estimate, then he engages the target. This continues until the target is destroyed or all rounds are expended (Figure 6-6).

Figure 6-6. Pair firing.

d. **Volley Firing**. Two or more firers can engage a single target when the range is known. These firers engage the target at one time on a prearranged signal such as a command, whistle, booby trap, mine, or TRP. For the best method of engagement, use the light antiarmor weapon, because it places the most possible rounds on one target at one time, increasing the possibility of a kill (Figure 6-7).

Figure 6-7. Volley firing.

e. **Communications**. Leaders control all unit fire and communicate this information to the entire unit IAW unit SOP. Light antiarmor weapons firers must know—

- Designated firers.
- Target priority.
- Method of engagement.
- Range and lead to target (if known).
- Command or signal to fire.
- Command or signal to cease fire.

## 6-5. ENGAGEMENT OF FIELD FORTIFICATIONS AND BUILDINGS

Light antiarmor weapons have little effect against field fortifications and buildings. Soldiers should not expect to severely damage targets with these weapons. However, if the alternatives shown in Table 6-2, page 6-8, are used, soldiers may be able to gain a temporary advantage.

## 6-6. ENGAGEMENT OF OTHER VEHICLES

The M72-series LAW proves more effective against light vehicles; the M136 AT4 proves more effective against armored vehicles. Nonarmored vehicles, such as trucks, cars, and boats, are considered *soft targets*. Firing along their length offers the greatest chance of a kill, because this type of shot is most likely to hit their engine block or fuel tank.

| AIM POINT | EFFECT WHEN WEAPON IS FIRED AT AIM POINT | RECOMMENDED FIRING TECHNIQUE |
|---|---|---|
| **BUNKER OR FIGHTING POSITION** | | |
| Firing Port or Aperture | Rounds fired into firing ports or apertures are wasted: rounds detonate inside rear of position, causing little structural damage to the position or to the equipment or personnel within, unless they are hit directly. The AT4 produces less effect than the LAW. | Coordinate fire: Fire *light antiarmor weapons* at a point 6 to 12 inches from the front edge of the firing ports in the berm. Fire *small arms* at the bunker or position to prevent personnel within from returning light antiarmor fire. |
| Berm | Firing at the berm causes the round to detonate outside the fighting position or inside the berm, creating only a small hole in the berm, dust, or minor structural damage to the position, but no damage to personnel or equipment unless they are hit directly. The AT4 produces less effect than the LAW. | |
| Window | The round may travel completely through the structure before detonating; if not, it creates dust and causes minor structural damage to the rear wall, but little damage to personnel or equipment, unless they are hit directly. The AT4 produces *less* effect than the LAW. | Fire 6 to 12 inches from the sides or bottom of a window. Light antiarmor rounds explode on contact with brick or concrete, creating an opening whose size is determined by the type of round used. |
| Wall | The round detonates on contact, creating dust and causing a small hole and minor structural damage, but little damage to personnel or equipment, unless they are hit directly. The LAW may be used to create a loophole, which is a hole large enough to throw hand grenades through. The AT4 produces *less* effect than the LAW. | |
| Corner | Corners are reinforced and thus harder to penetrate than other parts of a wall. Any light antiarmor round will detonate sooner on a corner than on a less dense surface. Detonation should occur in the targeted room, creating dust and causing *overpressure*, which can temporarily incapacitate personnel inside the structure near the point of detonation. The AT4 causes *more* overpressure than the LAW. | |

**Table 6-2. Effects of light antiarmor weapons on field fortifications or bunkers.**

## 6-7. LIMITED VISIBILITY ENGAGEMENTS

Limited visibility engagements can be conducted using various night vision devices or with artificial illumination. However, when NVDs or artificial illumination is used, limited visibility can reduce the maximum effective range for light antiarmor weapons by at least one-third. To avoid fratricide, leaders must ensure all designated light antiarmor weapon firers are trained to use their weapons in limited visibility.

a. **Night Vision Device**. Before a NVD can be used with the M136/AT4, it must be removed from its designated weapon (M249 machine gun or automatic weapon, or M60 machine gun) and the M136/AT4 mounting bracket must be attached. Appendix F describes the various NVDs that can be used, it also gives information for mounting, boresighting, and zeroing procedures for each NVD.

b. **Artificial Illumination**. If artificial illumination is used during a limited visibility engagement, it should be placed above and slightly beyond the target. However, the ability to identify and engage targets is even less with artificial illumination than with NVDs.

## 6-8. ENGAGEMENT IN NBC CONDITIONS

Wearing a protective mask limits the firer's ability to sight the weapon. Wearing NBC gloves limits his ability to manipulate the firing mechanism.

a. **Sighting the Weapon**. Sighting while wearing the protective mask may require rotating the weapon slightly counterclockwise. The mask also makes determining the location, identity, and engageability of targets more difficult.

b. **Firing the Weapon**. Practice manipulating the firing mechanism while wearing NBC gloves.

**NOTE**: When live firing either a light antiarmor weapon or its subcaliber trainer, aim within range firing limits.

> **DANGER**
>
> **THE M136 AT4 IS NOT RATED SAFE. IN TRAINING, NEVER FIRE IT FROM AN ENCLOSURE; IN COMBAT, DO SO ONLY WHEN NO OTHER POSITION IS AVAILABLE. THE OVERPRESSURE AND BLAST COMBINED CREATE CONDITIONS THAT CAN KILL YOU. IF THEY DON'T KILL YOU, THEY CAN INJURE YOU SERIOUSLY OR DEAFEN YOU TEMPORARILY OR PERMANENTLY.**

## 6-9. ENGAGEMENT FROM AN ENCLOSURE

Firing from an enclosure creates unique hazards. As such, before positioning soldiers in enclosures (in combat only), leaders must consider several factors that affect safety. Only in combat, *when no other tactical option exists*, should the M136 AT4 be fired from an enclosure. If it must be employed this way, the enclosure must meet the following minimum requirements. The M72-series LAW has been rated as safe for use from an enclosure but, again, only when the enclosure meets the following minimum requirements:

a. **Construction**. The building must be sturdily constructed to reduce the structural damage that would occur in a weakly constructed enclosure such as one made of wood or stucco.

b. **Size of Enclosure**. Minimum measurements for the enclosure are as follows:

- AT4—minimum room size 17 x 24 feet.
- LAW—minimum room size 12 x 15 feet.
- Both—minimum ceiling height 8 feet.

c. **Ventilation to the Rear and Sides**. To allow for the backblast, at least 20 square feet of ventilation—such as a standard 3-foot by 7-foot doorway—must be provided directly behind the firer. More doors and windows are removed beside and behind the position to increase ventilation and reduce overpressure, noise, and blast effects. Without sufficient ventilation, the blast would weaken or collapse the walls. On the front wall, windows and doors are reinforced rather than removed, because removing would draw attention to the position. Reinforcing the windows also helps protect the firer from enemy direct-fire weapons.

d. **Objects and Debris**. All objects and debris are removed from the rear of the weapon, because the backblast causes them to fly around the room injuring personnel.

e. **Muzzle Clearance**. Muzzle Clearance must be at least 6 inches.

f. **Weapon Clearance**. Properly positioning the weapons within the enclosure is vital to the safety and survival of all personnel in the enclosure. The weapons should be positioned so that no walls are within 5 meters to the rear or side of the weapon.

g. **Personnel Positions**. If any other soldiers must be present, then they must remain forward of the rear of the launcher and avoid standing in corners or near walls. If possible, they should construct reinforced positions that fit the previous criteria and that can protect them in case the building collapses.

**WARNING**

**To avoid injuring the eardrums, soldiers must wear the approved brand of ear protection.**

## 6-10. ENGAGEMENT BEYOND MAXIMUM EFFECTIVE RANGE (M136 AT4 ONLY)

A skilled M136 AT4 firer can engage targets beyond the weapon's maximum effective range of 300 meters, up to 550 meters. Beyond 550 meters, the firer must aim higher than center of mass and apply additional lead for moving targets. Commanders must realize that accuracy is reduced at these ranges. Also, firing at these ranges identifies the firing position to the enemy.

## 6-11. OFFENSIVE OPERATIONS

All elements, even those with other organic antiarmor weapons, use light antiarmor weapons. Light antiarmor weapons can influence the action in an attack, so units should routinely stock them beforehand. They are most useful against lightly armored vehicles. They can also be used against soft targets, such as bunkers, field fortifications, automobiles, and trucks, but their shaped-charge warheads have less effect on these than on armored targets. Unless personnel, ordnance, or flammable material on or inside them are hit, soft targets can normally continue to fight after being attacked by light antiarmor weapons. Due to their relatively short range, light antiarmor weapons should be placed throughout the attacking force. They support the maneuver by providing a base of fire, and they enable the assaulting force to engage in close antiarmor combat.

## 6-12. DEFENSIVE OPERATIONS

Whether or not other organic antiarmor weapons are available, light antiarmor weapons are an asset in the squad and platoon defensive plan.

a. Light antiarmor weapons are employed with interlocking fires to provide mutual support. Dispersion allows leaders the flexibility to place flank, rear, and oblique fires on targets. This procedure increases the survivability of the firers as well as the probability of achieving kills. Leaders must select positions that avoid fratricide from antiarmor backblast and short rounds (Appendix A).

b. The squad and platoon leader's sector sketch identifies TRPs and primary areas of possible engagement. The sector sketch also identifies possible avenues of approach for enemy armored vehicles. Each light antiarmor weapon shares a sector of fire with the primary small-arms weapon assigned to its position. However, because it is neither a primary nor crew-served weapon, the light antiarmor weapon does not require a separate range card. When assigning sectors of fire, squad and platoon leaders inform firers of all possible target areas, TRPs, and prearranged signals (Figure 6-8). (FMs 7-7, 7-7J, and 7-8 provide more information about the squad and platoon sector sketch.) Leaders provide each designated position with two or more light antiarmor weapons.

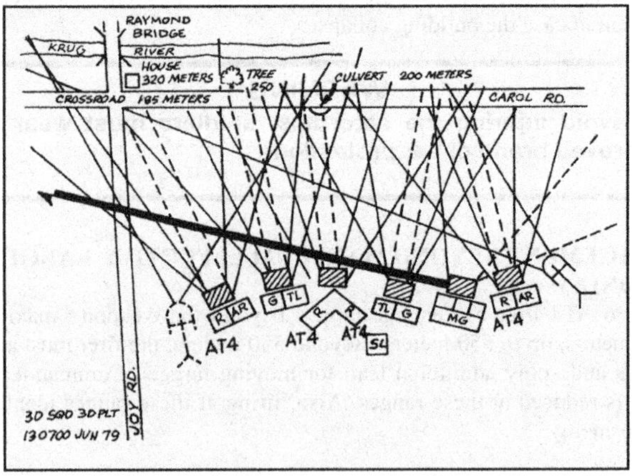


Figure 6-8. Squad sector sketch.

## 6-13. OTHER TACTICAL OPERATIONS

The weight of the light antiarmor weapon suits it well for combat patrols and rear area operations. (FM 7-8 provides more information about these subjects.)

a. **Combat Patrols**. Light antiarmor weapons are used on combat patrols to destroy enemy equipment, installations, and key points, and to harass enemy forces. The two types of combat patrols are—

(1) ***Ambushes (Antiarmor)***. The commander can employ armor-killer teams either during limited visibility or when cover, concealment, and withdrawal routes are available. The key to ambushing armored or other vehicles is to choose terrain that restricts their maneuverability and fields of fire, but that allows friendly forces to engage the vehicles from the flank and rear. Soldiers can emplace antipersonnel mines before the ambush so that dismounting enemy soldiers deploy into them. Though light antiarmor weapons can be used independently, they are normally used in support of designated organic antiarmor weapons such as the Dragon. Volley firing light antiarmor weapons increases the probability of a quick kill (Figure 6-9). Security teams stop the enemy from escaping.

**Figure 6-9. Antiarmor ambushes.**

(2) ***Raids***. Using light antiarmor weapons on raids increases target options. Light antiarmor weapons can also be used to defend against armored vehicles.

b. **Rear Area Operations**. Rear area operations usually consist of either: defending unit trains and tactical operation centers, or patrolling rear areas.

(1) ***Defense of Tactical Operations Center and Unit Trains***. The mobility of modern vehicles makes the TOC and unit trains prime targets. Light antiarmor weapons are used to engage vehicles threatening the unit during the defense of the TOC. Soldiers who use light antiarmor weapons in this type of operation normally perform noninfantry-type roles. Unit leaders designate soldiers to carry light antiarmor weapons and ensure that these soldiers receive light antiarmor weapons training.

(2) ***Patrols of Rear Areas***. Rear area patrols are security patrols conducted by designated infantry or MP units. These patrols can react to any threat they encounter in the rear area. Rear area patrols use the light antiarmor weapon in a hasty point defense at a roadblock, intersection, or strongpoint.

CHAPTER 7

# TRAIN-THE-TRAINER AND UNIT TRAINING PROGRAMS

*During initial-entry training, soldiers receive only an orientation for each of the light antiarmor weapons. When they arrive in a unit, soldiers should train to Skill Level 1 proficiency on all assigned weapons and equipment. Each unit should have an ongoing unit training program to allow soldiers to work together to attain and sustain their skills. This program integrates individual and collective tasks. Operating a unit training program requires the unit to also maintain a train-the-trainer program. Both programs are planned and conducted IAW the commander's METL.*

## 7-1. TRAINING STRATEGY

Training strategy integrates resources into a year-round program to train the individual and collective skills needed to perform the wartime mission. Thus, units are trained to fight and win on the battlefield. The training strategy includes institutional and unit training.

a. **Institutional Training**. TRADOC schools conduct institutional training. These courses train critical individual and leader skills required to succeed in combat. Students are provided the latest doctrine and taught current technical skills associated with the light antiarmor weapons available within the Army. Soldiers train at several institutional schools throughout their career (see the appropriate programs of instruction [POIs]) (Table 7-1).

| | |
|---|---|
| **ENLISTED** | Initial-Entry Training |
| | Primary Leadership Development Course |
| | Basic Noncommissioned Officer Course |
| | Advanced Noncommissioned Officer Course |
| **OFFICERS** | Precommission Training |
| | Officer Basic Course |
| | Officer Advanced Course |

**Table 7-1. Institutional training.**

b. **Unit Training**. Unit training consists of sustainment training in individual and collective skills. Commanders manage their resources (devices, simulators, simulations, ranges, and ammunition) carefully to best develop individual and leader skills. Through the use of exercises, (drills, STXs, and qualification tables), the unit develops skills and integrates individuals into a cohesive crew or squad. Table 7-2 provides a training guide for building a light antiarmor weapons sustainment training plan (see also Appendixes C and E).

## 7-2. UNIT TRAINING PROGRAM

Every unit commander, including those commanding CS and CSS units, should develop a program for light antiarmor training. Some units have more than one type of light antiarmor weapon that requires sustainment training. The commander considers preparation and training for both collective and individual tasks. Trainers analyze their collective tasks and the individual tasks that support them. Individual tasks must be integrated into collective training and rehearsals. The units should try to train light antiarmor weapons at the same time and on the same range (Appendix D). Otherwise, these weapons are trained in alternating quarters of the year.

a. **Sustainment Training Requirements**. Sustainment training is recommended for each soldier. Active Component units should conduct sustainment training quarterly and fire the appropriate qualification tables semiannually, depending on the STRAC TRC level. Reserve Component units should conduct sustainment training semiannually and fire the appropriate instructional or qualification tables annually. All units should fire advanced instructional tables semiannually. When possible, they should conduct live fire annually IAW the current STRAC manual. (Appendix E provides all the necessary firing tables.) Commanders should ensure that designated gunners complete unit sustainment training and live fire tactical light antiarmor weapons sometime during that quarter. *Training requirements* are divided into four tracks, A through D, based on ammunition allocations (Table 7-3). Table 7-4 provides an example *training frequency chart*. This example is based on the AT4. The same tasks apply to all light antiarmor weapons, but procedures and standards may vary. Tactical weapons are allocated IAW the current standards in training commission (STRAC) manual.

| | |
|---|---|
| 1 | Designated soldiers should train as often as needed to maintain proficiency. |
| 2 | Performance evaluations should be administered to designated soldiers quarterly, if resources allow. |
| 3 | Designated soldiers should fire the qualification tables semiannually for verification on designated weapons. |
| 4 | Designated soldiers qualify semiannually on their designated weapons. Qualification consists of a hands-on performance evaluation and qualification firing tables. |
| 5 | Designated soldiers should be trained in advanced *instructional* techniques as often as the unit training program allows. This training helps to maintain their target engagement skills. |
| 6 | All soldiers not designated as light antiarmor weapons gunners should train on these weapons to the extent allowed by resources. |

**Table 7-2. Sustainment training plan for light antiarmor weapons.**

| TRACK | DETERMINES ANNUAL AMMUNITION ALLOCATION FOR: |
|---|---|
| A | Each Active Component designated firer. |
| B | Track deleted |
| C | Each Reserve Component designated firer. |
| D | Each Reserve Component training division instructor. |

**Table 7-3. Training requirements.**

| TRAINING OBJECTIVE | TASK | FREQUENCY | | APPROXIMATE TIME |
|---|---|---|---|---|
| | | AC | RC | |
| NA | Conduct an orientation safety briefing (as appropriate). | Q | A | 10 min |
| 1 | Inspect an M136 AT4 for serviceability. | Q | A | 30 min |
| 2 | Prepare an M136 AT4 for firing. | Q | A | 45 min |
| 3 | Demonstrate correct M136 AT4 firing positions. | Q | A | 45 min |
| 4 | Estimate range. | Q | A | 60 min |
| 5 | Determine the correct M136 AT4 sight picture. | Q | A | 30 min |
| 6 | Perform M136 AT4 misfire procedures. | Q | A | 45 min |
| 7 | Return the M136 AT4 to the carrying position. | Q | A | 45 min |
| 8 | Engage targets with an M136 AT4. | S | B | 3 1/2 hr |
| **KEY:**<br>**Q = quarterly (every 90 days)**<br>**S = semiannually (twice a year)**<br>**A = annually (once a year)**<br>**B = biennially (every two years)** | | | | |

**Table 7-4. Example training frequency chart.**

b. **Gunnery Training Requirements**. All designated soldiers should be trained to standard IAW the combat techniques in Chapter 6 and the performance evaluations in Appendix C.

c. **Collective Training**. Individual tasks must be integrated into collective training and rehearsals. To accomplish this, commanders analyze the collective tasks from their unit METL and the individual tasks that support the collective tasks. The commander determines the collective tasks that must be practiced by analyzing the mission outlines in the appropriate mission training plan (MTP). The mission outlines describe the collective tasks required to execute each ARTEP mission. Leaders determine the individual tasks that support the collective tasks by referring to the mission task matrix in the appropriate MTP.

(1) Leaders use the training standards given in the appropriate manual for the individual tasks. Noninfantry units can use FMs 7-7, 7-7J, 7-8, 7-10, and 7-20 to identify infantry-type missions. Some of the tasks identified may include engaging armored vehicles. When planning training, the trainer incorporates the training devices and aids for light antiarmor weapons into the unit's training. These devices and aids add realism and allow trainer to properly evaluate the task being conducted. He must also identify deficiencies and correct them by retraining soldiers to use the selected weapons and their related training devices correctly.

(2) The MILES Viper is a force-on-force trainer for light antiarmor weapons. MILES trainers are *not* gunnery trainers and are not used for sustainment training. They are used in all force-on-force training. Table 7-5 provides an example MILES training program.

| SEQUENCE | FREQUENCY (TIMES PER YEAR) | |
|---|---|---|
| | ACTIVE | RESERVE |
| Squad, platoon, or company FTX or STX | 4 | 1 |
| Battalion FTX | 2 | 0 |
| Battalion ARTEP | 2 | 1 |
| Total MILES training events per year | 8 | 2 |

**Table 7-5. MILES training program.**

## 7-3. TRAIN-THE-TRAINER PROGRAM

The success of the unit's light antiarmor training program depends on trainers who are well-trained, competent, and prepared to train soldiers.

a. **Objectives**. The train-the-trainer program must teach trainers to do the following:

- Evaluate the unit's training weaknesses.
- Plan training.
- Set up, operate, and maintain training equipment.
- Conduct training.
- Coach firers in gunnery training.
- Evaluate firers' training weaknesses.
- Correct firers' training weaknesses.
- Instill confidence in the firers trained.
- Maintain training records.

b. **Mission-Essential Task List**. The commander must carefully examine his unit's wartime mission and develop a METL. He must then organize the unit's sustainment training program to support the METL. The light antiarmor sustainment training plan consists of quarterly, semiannual, and annual training events. Between these events, the unit provides additional light antiarmor weapons training. The sustainment plan should prepare the unit for the next round of weapons gunnery while also supporting the unit METL.

c. **Trainer Assessment**. Trainers within a unit are normally team, squad, and section leaders and platoon sergeants. Before becoming weapons trainers, they must be assessed carefully and any shortcomings corrected. The commander chooses a method of assessing the trainers that ensures that their abilities are accurately evaluated. Then, with the assistance of the platoon leaders and platoon sergeants, the commander performs this assessment.

d. **Command Benefits**. Regular and aggressive train-the-trainer programs provide valuable benefits that gradually spread through the unit. These benefits include a base of expertise, proficiency, and esprit de corps. Maintaining this base is easier than recreating it.

e. **Training Tasks**. Successful trainers know how to operate the training devices for the light antiarmor weapons assigned to their units. Based on the unit's METL, the commander designates the appropriate light antiarmor weapon for specified missions. Appendix C provides the tasks, conditions, and standards for each weapon discussed in this manual. The trainers must know the appropriate combat techniques for employing these weapons. They must also perform the following:

- Conduct an orientation safety briefing.
- Inspect the weapons for serviceability.
- Prepare the weapons for firing.
- Demonstrate the correct firing positions.
- Estimate range.
- Determine the correct sight picture.
- Perform the correct combat and training misfire procedures.
- Return the weapon to the carrying configuration.

# APPENDIX A
# SAFETY

*This appendix provides safety procedures for both training and combat.*

## A-1. TRAINING SAFETY

Training safety for light antiarmor weapons applies to all types of ranges and training sites.

a. **Backblast**. Light antiarmor weapons have little or no recoil, because the propellant gasses escape to the rear of the weapon. This backblast can damage equipment or seriously injure personnel who are too close to the rear of the launcher. When operating temperatures fall below freezing (0NC or 32NF), all backblast areas and safety zones double.

(1) ***M72-series LAW***. This is an open-chambered weapon, so it has no recoil. The launcher's total backblast area extends 40 meters (44 yards) to the rear (Figure A-1) and is divided into two zones. During training, both should be marked off limits:

(a) Danger zone. All personnel, equipment, and flammable material must be clear of this area.

(b) Caution zone. The weapon's backblast may throw loose objects to the rear. Therefore, personnel must also stay clear of this area.

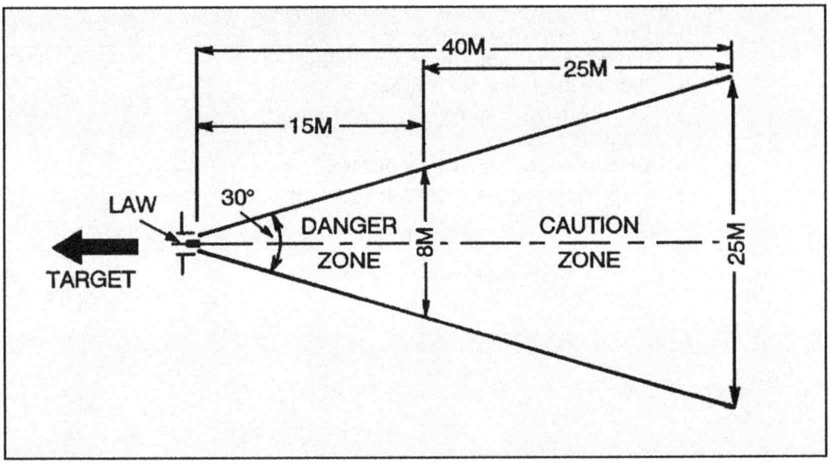


**Figure A-1. M72-series LAW backblast area.**

(2) ***M136 AT4***. Because the AT4 has a closed chamber, it has a slight recoil. When the weapon is fired, pressure builds up in the chamber and ruptures the baseplate. Once the baseplate ruptures, gasses exit from the back of the weapon, forming the backblast. The total backblast area extends 100 meters to the rear of the launcher in a 90-degree fan (Figure A-2, page A-2). During training, the *entire* backblast area should be marked off limits and kept clear of personnel and equipment. In combat, personnel may remain to the rear of the

weapon, but *only* if they are in protected positions. Backblast heat or flying debris could injure the firer and other soldiers anywhere in the backblast area. This weapon must not be fired from an enclosure or in front of a barrier (TM 9-1315-886-12). It may be fired from an enclosure in combat *only when* no other tactical option exists.

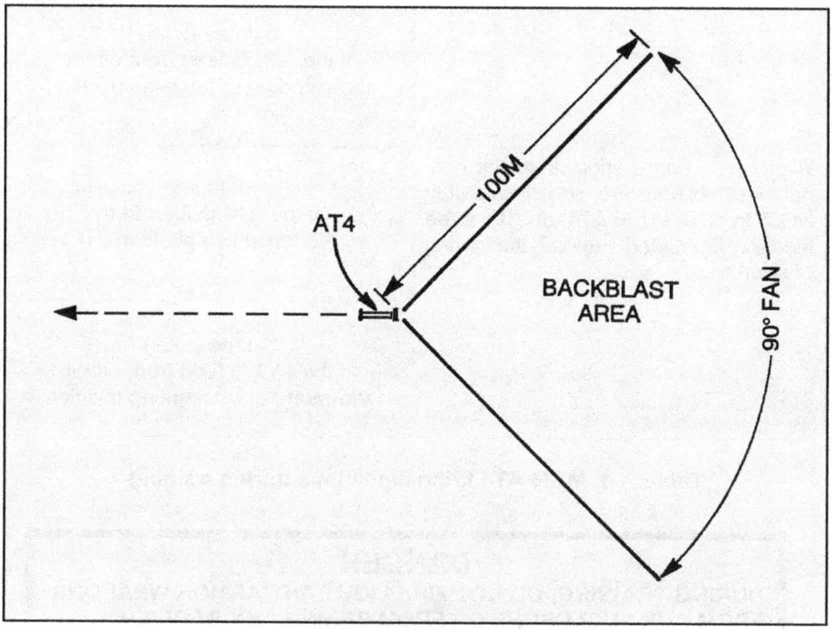


**Figure A-2. M136 AT4 backblast area.**

b. **Firing Limitations**. Trainers and soldiers must observe the backblast danger zones as well as other firing limitations. A light antiarmor weapon can be fired from the standard infantry fighting position. However, if this position is chosen, trainers/soldiers must consider other factors that increase the accuracy and reduce the danger to friendly soldiers: Rear of launcher extends beyond the back wall of the position, no overhead cover extends into the backblast area, and at least 1/2 meter (18 inches) clearance is around the weapon.

c. **Range Procedures for Light Antiarmor Weapons**. For all the light antiarmor weapons discussed in this manual, these procedures must be followed:

(1) The whole backblast area must be well marked and must be clear of all personnel, equipment, and flammable materials.

(2) No one enters the area behind the firing line or forward of the rear safety line without permission from the range OIC or safety NCO.

(3) Weapons are stored to prevent exposure to the sun or extreme temperatures.

d. **Additional Range Procedures for the M136 AT4**. The extreme overpressure and noise created when this weapon is fired require special precautions during training. Not only must firers and others present wear earplugs, but they may only use the approved brand—no other brand provides sufficient protection. To reduce these hazards, the range firing limitations shown in Table A-1 must be followed.

| | |
|---|---|
| Within a 24-hour period, a soldier may only fire, observe fire, or act as safety NCO for the M136 AT4 one to three times, depending upon the firing position used | 3 times (total)<br>if the AT4 is fired from either<br>the standing or kneeling position<br>OR<br>2 times<br>if the AT4 is fired from<br>the sitting position<br>OR<br>1 time (total)<br>if the AT4 is fired from either<br>the prone or the fighting position |

**Table A-1. M136 AT4 firing limitations during training.**

**DANGER**
**DURING TRAINING, DO NOT FIRE LIGHT ANTIARMOR WEAPONS FROM ANY ENCLOSURE OR FROM BEHIND ANY BARRIER.**

## A-2. COMBAT SAFETY

Combat safety rules and procedures include all those that apply to training plus the following:

a. **Fighting Position**. Both the M72-series LAW and the M136 AT4 can be fired from the standard infantry fighting position. However, to increase accuracy and reduce the danger to friendly soldiers, the area to the rear of the firing position must have no walls, large trees, or other obstructions within 5 meters (5 1/2 yards) to deflect the backblast onto the firer or into the position.

**DANGER**

**ENSURE THE AREA TO THE REAR OF THE FIRING POSITION HAS NO WALLS, LARGE TREES, OR OTHER OBSTRUCTIONS WITIN 5 METERS (5 1/2 YARDS). OBSTRUCTIONS DEFLECT BACKBLAST ONTO THE FIRER OR INTO THE POSITION, INJURING OR KILLING THE FIRER AND ANY OTHER SOLDIER OCCUPYING THE FIRING POSITION.**

(1) In the *individual infantry fighting position*, the soldier must lean against the rear wall and ensure that the venturi or the rear of the weapon protrudes past the rear of the position. The firer's elbows are not supported.

(2) In the *two-soldier infantry fighting position*, the nonfiring soldier must remain clear of the backblast area. These positions should be constructed and sited so that none are located in another's backblast danger zone.

(3) A *modified firing position* may be constructed to the side of the two-soldier fighting position. Firing from a modified position reduces the possibility of injury to the firer or another soldier in the fighting position, but it offers the firer some protection from enemy return fire.

b. **Temperature Limitations**. No light antiarmor weapon should be fired when its temperature exceeds operating limits. When a weapon temperature is below freezing (0 degrees Centigrade or 32 degrees Fahrenheit), overpressure increases and prevents some of the propellant from burning. Then, as the projectile leaves the launch tube, portions of the remaining propellant could be directed onto the firer's face or upper body. Consequently, in cold weather, the firer should wear a protective mask. On the other hand, when the weapon temperature is above 60 degrees Centigrade (140 degrees Fahrenheit), propellants and warhead compounds can destabilize, so, whether they are located at training sites or fighting positions, weapons should be stored in a shaded area in hot weather. Finally, weapons must not be fired when their temperatures exceed safe operating limits: -40 degrees to 60 degrees Centigrade (-40 degrees to +140 degrees Fahrenheit).

c. **Minimum Target Engagement Ranges**. In combat, no targets will be engaged within the 10-meter minimum arming range. During training, no targets will be engaged within 30 meters.

d. **Overhead Fire**. Light antiarmor weapons must not be fired over the heads of friendly soldiers, unless the soldiers have adequate protection against direct impact or other hazards.

APPENDIX B
# TRAINING DEVICES AND AIDS

*Training devices and training aids enable soldiers to learn as much as they can about a weapon before they try the real thing. This procedure saves money and time, and it prevents injuries.* ***Training devices*** *both look and behave like the real weapon, but are cheaper and safer to practice with.* ***Training aids*** *include anything else used to help soldiers learn to use a weapon. Trainers should use their imaginations and invent or adapt other training aids from available resources.*

## B-1. LAW SUBCALIBER TRAINING DEVICE

The only M72-series LAW training device designed for individual use is a subcaliber trainer (Figure B-1). This device, referred to as the M190 subcaliber launcher, is made by adding an M190 subcaliber conversion kit (NSN 1340-00-420-7999) (Figure B-2) to an expended M72-series LAW launcher. It is used to fire the M73 subcaliber (35-mm) rocket. This 36-mm rocket is shorter and lighter than the LAW's 66-mm tactical rocket. It simulates the tactical rocket's smoke and flight trajectory but with less noise and backblast. The expended launcher is likely to wear out before the subcaliber inner tube does. When this happens, the inner tube can be transferred to another expended launcher. The M190 subcaliber launcher can be used in all training phases, from a fixed firing line to simulated tactical situations such as a squad live-fire exercise.

**NOTE**: The local Training Support Center (TSC) can install the conversion kits, but DOD Regulation 5100.76-M requires that all sensitive conventional arms, ammunitions, and explosives, including expended launchers, be carefully controlled.

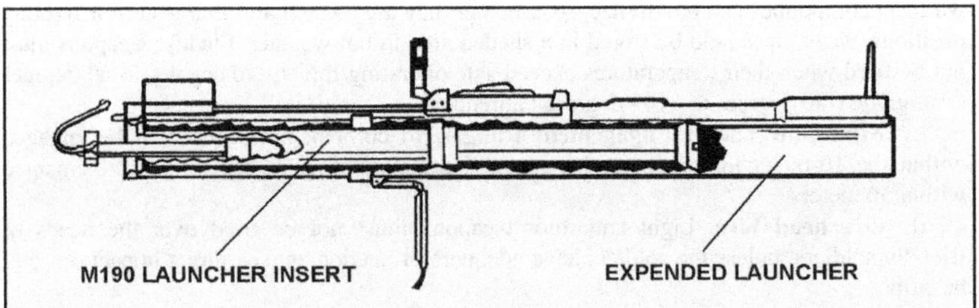


**Figure B-1. LAW subcaliber trainer.**

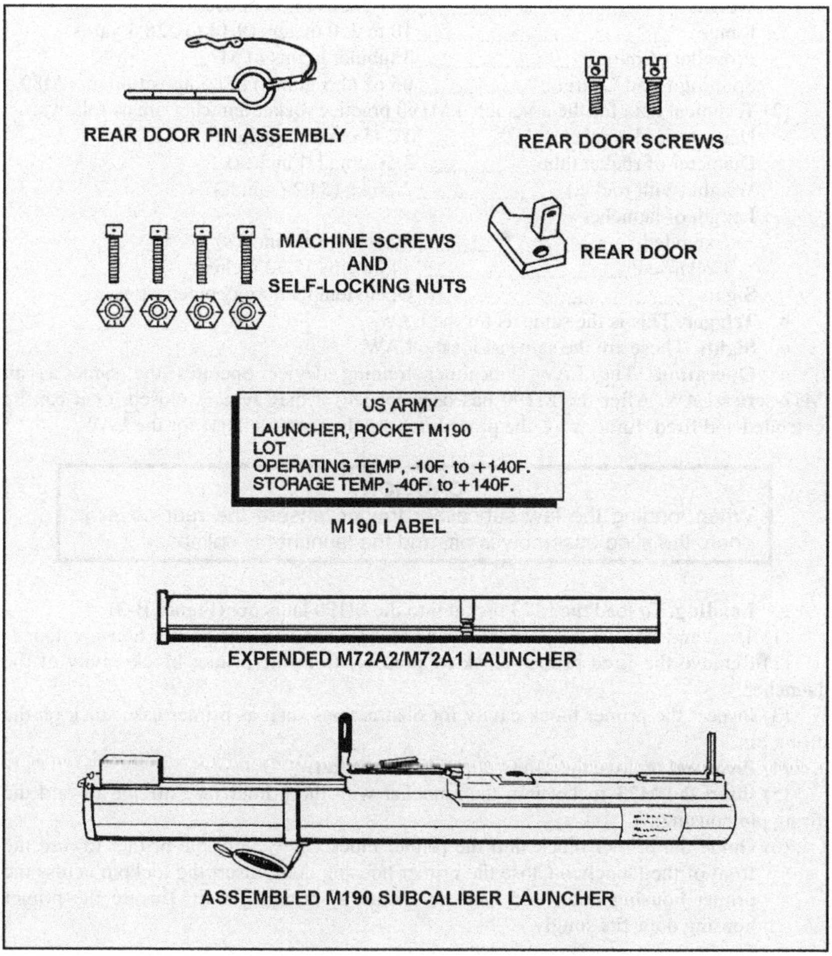


**Figure B-2. M190 subcaliber conversion kit.**

a. **Technical Data.**

(1) Technical data for the M73 35-mm subcaliber practice rocket are as follows:

| | |
|---|---|
| Length | 22.48 cm (8.86 inches) |
| Diameter | 3.51 cm (1.38 inches) |
| Weight | 154 grams (5.39 ounces) |
| Range | 10 to 250 meters (9.14 to 228.5 yards) |
| Propellant Charge | 3 tubular grains of M7 |
| Spotting Head Charge | 05 oz (1.5 grams) of composition mix M80 |

(2) Technical data for the assembled M190 practice rocket launcher are as follows:

| | |
|---|---|
| Diameter of launcher | 12.45 cm (4.9 inches) |
| Diameter of rocket tube | 3.56 cm (1.4 inches) |
| Weight (with rocket) | 2.26 kg (5.02 pounds) |
| Length of launcher | |
| Extended | 89.9 cm (35.4 inches) |
| Collapsed | 64.26 cm (25.32 inches) |
| Sights | Open, temperature compensating |

b. **Trigger.** This is the same as for the LAW.

c. **Sights.** These are the same as for the LAW.

d. **Operation.** The LAW subcaliber training device operates the same as an M72-series LAW. After the M190 has been loaded, it may remain closed, or it can be extended and fired. Either way, the procedures are the same as those for the LAW.

**CAUTION**

When loading the law subcaliber trainer, ensure the rear cover is open, the sling assembly is off, and the launcher is collapsed.

e. **Loading.** To load the M73 rocket into the M190 launcher (Figure B-3)—

(1) Twist and remove the primer housing lockpin and open the primer housing door.

(2) Remove the fired primer block (if present) from the primer block cavity of the launcher.

(3) Inspect the primer block cavity for obstructions such as primer case stuck on the firing pin.

(4) Pivot and remove the safety clip from the rocket.

(5) Insert the M73 rocket into the launcher with the primer face up and toward the firing pin housing.

(6) Guide the primer block into the primer block cavity with the primer toward the front of the launcher. Close the primer housing door, insert the lockpin across the primer housing door, and twist to ensure a spring-tight fit. Ensure the primer housing door fits snugly.

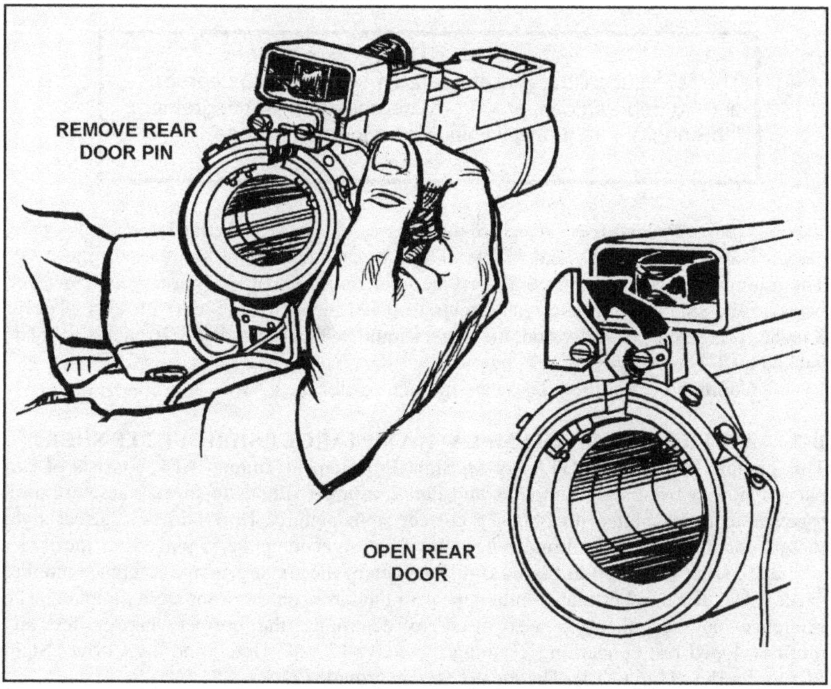


**Figure B-3. Loading the LAW subcaliber launcher.**

f. **Unloading**. To unload the M73 rocket from the M190 launcher—

(1) Return the arming handle to SAFE.

(2) Remove the weapon from your shoulder, keeping the weapon pointed downrange.

(3) Partly collapse the launcher, and remove the primer housing lockpin.

(4) Open the primary housing door, and remove the primer from its housing.

(5) Remove the rocket.

(6) Replace the safety clip on the rocket.

(7) Place the rocket in its original container.

g. **Function**. The M73 rocket launches the same as tactical rocket, but when the rocket head strikes the target, it sets off the spotting head that produces a flash, noise, and white smoke.

> **CAUTION**
> The M73 subcaliber rocket operates between –23 degrees and 60 degrees centigrade (-10 degrees and 140 degrees fahrenheit). Fire it only within this temperature range.

h. **Misfire Procedures**. These are the same as for the M-72 series LAW.

i. **Range Procedures and Safety**. These procedures are the same as for the LAW. The training device can be used against all solid stationary or moving targets. However, because the subcaliber rocket can penetrate 0.124 inches (0.315 cm) of steel plate or 8 inches (20.32 cm) of soft wood, the target should be constructed of 3/16-inch steel plate backed by 3/4-inch plywood.

j. **Combat Techniques**. These are the same as for the LAW.

## B-2. AT4 FRONT SIGHT TEMPLATE AND TARGET SILHOUETTE SHEET

This graphic training aid (GTA 7-2-5), Sight Engagement Trainer, AT4, consists of two parts: Part 1, a front sight template, and Part 2, a target silhouette sheet. These are used together to train soldiers to obtain a correct sight picture. To obtain the correct sight picture for the target speed and range, the front sight template is placed on the target silhouette sheet as shown in Figure B-4. For each of the six targets, the reference number sheet shows the number that should appear in the circle on the front sight template. The reference number sheet is also used to determine the correct answer for any unit-developed test concerning Training Objective 12 task, Determine the Correct Sight Picture for the M136 (AT4). The local TSC can provide GTA 7-2-5.

**Figure B-4. Using GTA-7-2-5 with target silhouette sheet and front sight template.**

## B-3. AT4 FIELD HANDLING TRAINER

The AT4 field handling trainer (FHT) is an inert device made from an expended AT4 tactical launcher. It simulates the weight, balance, characteristics, and operation of the launcher. The FHT is marked with either a gold or yellow 1-inch band between the front and rear sights, and with the word "DUMMY" in 1-inch letters on the side of the launch tube. The local TSC can convert expended launchers to training devices, but DOD Regulation 5100.76-M requires that these items be carefully controlled. The FHT is appropriate for several training situations that do not require the M287 tracer trainer, which prevents unnecessary damage to the tracer trainer. The FHT is used in marksmanship training, FTXs, and institutional training to help soldiers learn how to do the following:

- Inspect the AT4 for serviceability.
- Prepare an AT4 for firing.
- Demonstrate the correct AT4 firing position.
- Perform misfire procedures.
- Return an AT4 to carrying position.

## B-4. AT4 SUBCALIBER TRACER TRAINER

The M287 subcaliber tracer trainer (Figure B-5) uses the 9-mm M939 training practice-tracer (TP-T) cartridge. When loaded, this trainer simulates the AT4 in weight, balance, and operation. The velocity and trajectory of its ammunition match that of the AT4's HEAT cartridge, but the M287 produces less noise, backblast, and overpressure. The M287 subcaliber tracer trainer is used in place of the AT4 in training.

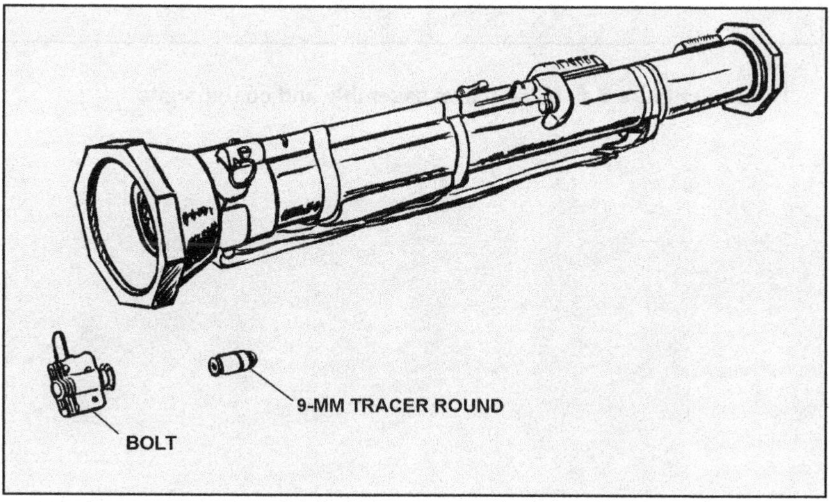


**Figure B-5. M287 subcaliber tracer trainer.**

a. **Description**. The M287 tracer trainer is a specially constructed AT4. Unlike the LAW subcaliber trainer, it is *not* made from an expended launcher, so it has its own NSN. It is designed to accept a special rifle barrel that fires a reduced-load 9-mm cartridge.

(1) The M287 may be fired at stationary or moving targets. Before it can be fired at a target tank occupied by personnel, the parts of the tank that could suffer damage must be shielded. The local TSC can provide specifications for modifying tanks to be used as targets for the M287 tracer trainer.

(2) Unlike the M136 AT4 itself and the FHT, the M287 has no band between the front and rear sights. The trainer has a 9-mm submachine gun barrel, a breach assembly, and a bolt (Figure B-6). The bolt is easily removed to load the 9-mm cartridges and to inspect the barrel for obstructions.

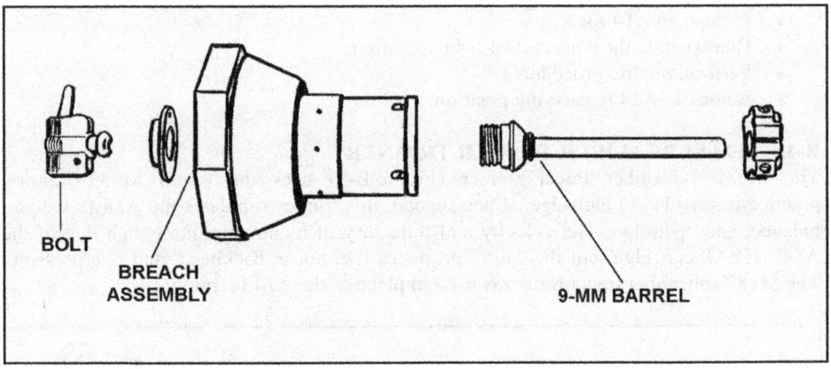


**Figure B-6. M287 barrel assembly and components.**

(3) If the M287 tracer trainer is damaged, the unit armorer can replace its complete firing mechanism, firing rod, bolt, sights, sling, and 9-mm barrel assembly (Figure B-7).

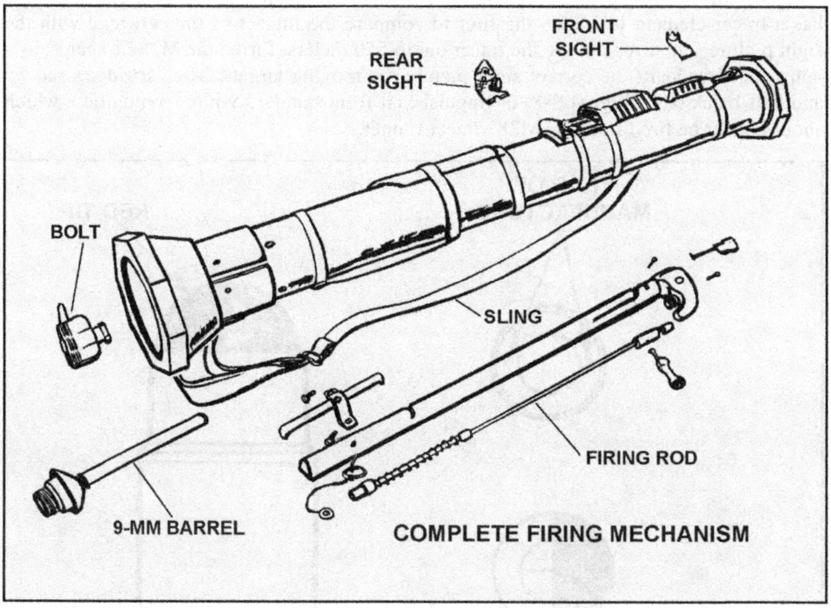


**Figure B-7. Replaceable parts for the M287 tracer trainer.**

b. **Technical Data**. The following technical data apply to the M287 subcaliber tracer trainer:

| | |
|---|---|
| Length | 1,020 mm (40 inches) |
| Weight | 7 kg (15 pounds) |
| Action | Mechanical |
| Sights: | |
| Front | Three-post system |
| Rear | Adjustable range, with two peepholes: 2-mm peephole for daylight and 7-mm peephole for limited visibility |
| Operating temperature | 10 degrees to 27 degrees Centigrade |
| | (0 degrees to 100 degrees Fahrenheit) |
| Muzzle velocity | 300 mps |
| | (984 fps) |
| Caliber | 9 mm |

c. **Ammunition**. The M939 9-mm tracer cartridge has a lighter powder charge than a standard 9-mm bullet. The lighter charge enables the cartridge to closely duplicate the trajectory of the AT4 tactical round at ranges out to 700 meters. The M939 cartridge also has a tracer element to enable the firer to compare the impact of the cartridge with the sight picture. The firer can see the tracer out to 550 meters. Firing the M287 tracer trainer helps the firer learn the correct sight picture for moving targets. The cartridge's red tip and half-black base (Figure B-8) distinguishes it from standard 9-mm ammunition, which should never be fired from the M287 tracer trainer.

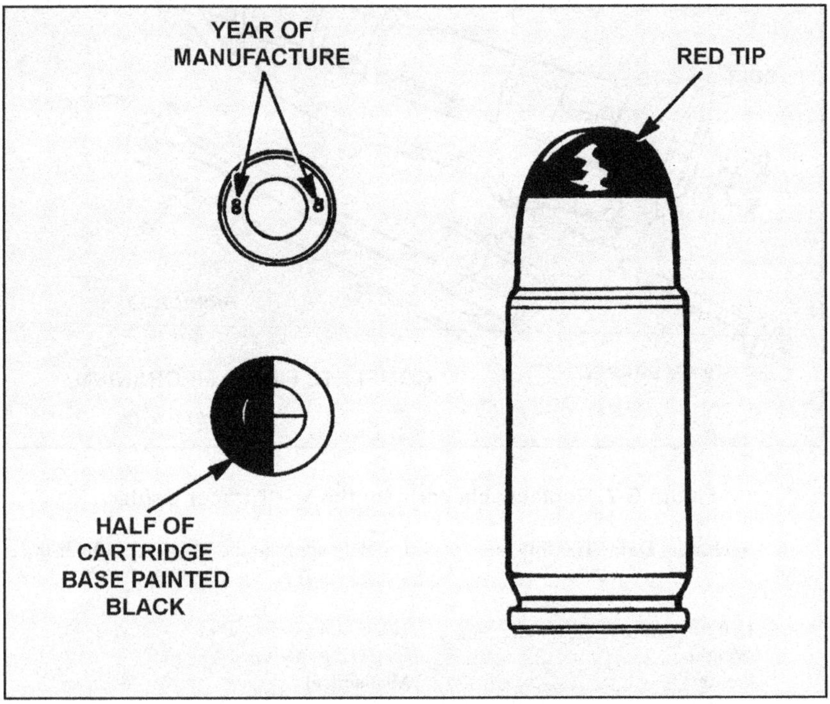


**Figure B-8. M939, 9-mm tracer cartridge.**

d. **Function Check**. Before the M287 is fired, a function check must be performed to ensure the trigger and safety mechanisms are operating properly. Function check procedures are shown in Table B-1. However, before performing a function check, ensure—

- The *cocking lever* is in the safe ("S") position.
- The *transport safety pin* is fully inserted, with the lanyard wrapped clockwise around the launcher.
- The *bolt* is removed from the breech.

| STEP | OPERATOR ACTIONS | FUNCTION CHECK | CORRECTIVE MEASURES |
|---|---|---|---|
| 1 | Try to cock the tracer trainer | It should not cock | If it cocks, turn the trainer in for repair |
| 2 | Remove the transport safety pin and depress the forward safety | The forward safety should spring back when released | If it does not spring back, turn the tracer trainer in for repair |
| 3 | Cock the firing mechanism, pressing only the red trigger button | The firing rod should move only slightly and should be captured by the safety lever assembly. No more than half the rod should protrude through the rear of the firing assembly | If the tracer trainer fires, turn it in for repair |
| 4 | Recock the firing mechanism; fully depress and continue to hold down the forward safety; press the red trigger button | The tracer trainer should fire | If the tracer trainer *fails* to fire, turn it in for repair |
| 5 | Return the cocking lever to the SAFE position and reinstall the transport safety pin | | |

**Table B-1. Function check, M287 tracer trainer.**

**WARNING**

**LOAD LIVE AMMUNITION ONLY ON THE FIRING LINE. NEVER FIRE THE M287 TRACER TRAINER AT HARD TARGETS LESS THAN 125 METERS FROM THE FIRING LINE. REMAIN CLEAR OF THE FRONT OF THE M287, WHICH MUST BE POINTED DOWNRANGE AT ALL TIMES.**

e. **Loading**. The M287 tracer trainer is loaded as follows:

(1) The firer performs the function check and places the M287 tracer trainer on his shoulder.

(2) The trainer looks through the barrel from the rear to verify that it contains no obstructions. He inspects the cartridge primer to ensure it is not dented, inserts the bolt into the breech (Figure B-9), and presses in and turns the bolt clockwise to the "S" to

place it in the safe position (Figure B-10). Then he arms the tracer trainer by turning the bolt clockwise to "F" so that it can be fired (Figure B-11).

(3) With the launcher still on his shoulder, the firer cocks the launcher by placing the cocking lever in the cocked position.

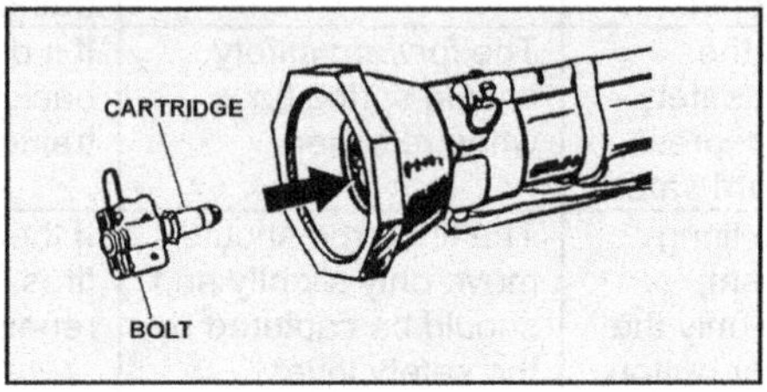


**Figure B-9. Inserting bolt with cartridge into the breech of the M287 tracer trainer.**

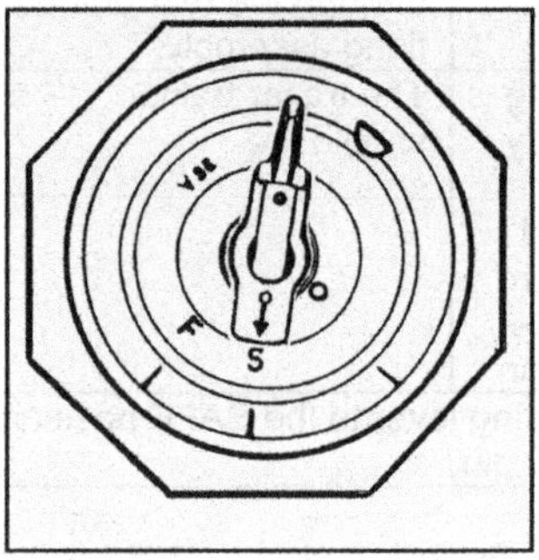


**Figure B-10. Ensuring tracer trainer bolt is in the safe position.**

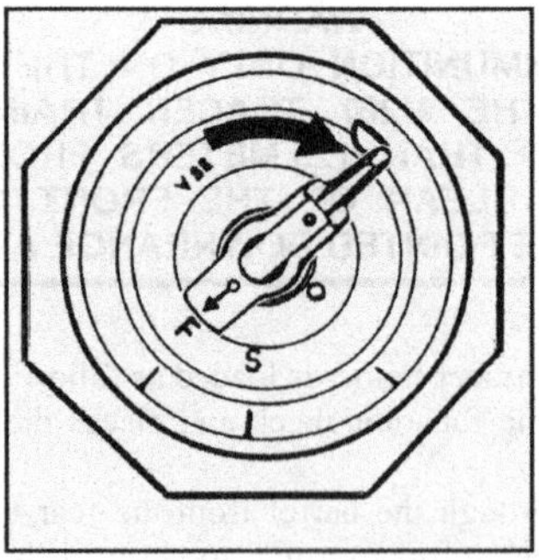


**Figure B-11. Arming the tracer trainer.**

f. **Unloading**. With the tracer trainer still on the firer's shoulder, the trainer or instructor removes the bolt by turning it counterclockwise past the "S," then pulling the bolt from the breech (Figure B-12). If more rounds are to be fired, he removes the expended cartridge and reloads the bolt.

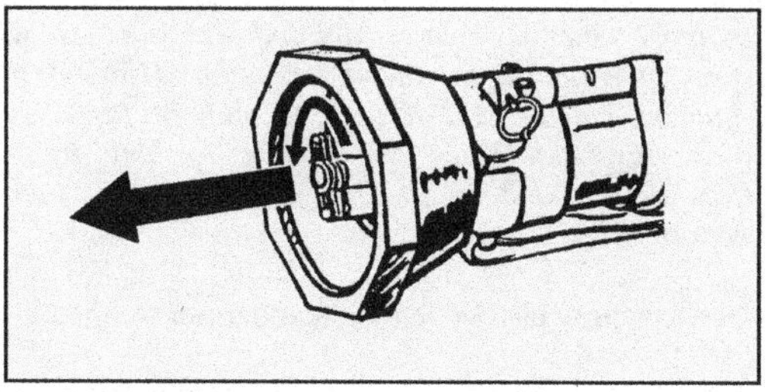

**Figure B-12. Unloading the tracer trainer.**

g. **Reloading**. With the tracer trainer still on his shoulder, the firer places the cocking lever in the safe position, keeping the tracer trainer on his shoulder and pointing toward the target. The trainer reloads the tracer trainer and prepares it for firing.

**DANGER**
**NEVER TOUCH THE TRIGGER WHILE RELOADING.**

h. **Misfire Procedures**. In the event of a misfire—

(1) Immediately shout "Misfire!"

(2) Release the forward safety, recock the launcher, and ensure the cocking lever is erect. Fully depress and hold down the forward safety, then aim and press the red trigger button.

(3) If the tracer trainer still fails to fire, repeat the two previous steps.

(4) If the launcher fails to fire again, release the forward safety, maintain your firing position, and return the cocking lever to the safe position.

(5) Notify the training supervisor.

(6) If a real misfire occurs that cannot be corrected by following the steps previously discussed, the trainer or assistant trainer on duty must ensure the cocking lever is in the safe position and that the forward safety is in the vertical position. He must then insert the transport safety pin and remove and inspect the cartridge. If the primer is dented, he replaces it, disposing of the old one IAW range safety SOP. If the primer is not dented, he notifies DS maintenance to inspect the bolt firing pin for damage. He then reloads the tracer trainer and continues the training exercise.

i. **Range Procedures and Safety**. These are the same as for the M136 AT4.

APPENDIX C

# PERFORMANCE EVALUATIONS

*The trainer administers performance evaluations to determine how much training the soldiers have learned. Those who fail are retrained and retested, and those who pass help retrain and evaluate those who did not. Instructions are given for right-handed firers: the M136 AT4 may only be fired right-handed, but the M72-series LAW may be fired left-handed, if procedures are reversed. Tasks are generic to both light antiarmor weapons. Conditions and standards may also be generic but, if not, are given by weapon.*

**TASK 1**: Perform serviceability checks on a light antiarmor weapon.

**CONDITIONS**: Given an expended light antiarmor weapon under suitable conditions or in an appropriate location.

### M72-Series LAW

**STANDARDS**: The firer performs the following actions:

### TRAINING AND EVALUATION

1. Inspects the launcher to ensure it has no cracks, dents, or bulges.

2. Ensures the trigger arming handle is present and in the SAFE position.

3. Checks the rubber boots around the trigger bar and detent for cracks, tears, and deterioration.

4. (*M72A2 only*) Inspects the data plate on the right forward side of the launcher to ensure the words "with coupler" appear on the first line. If not, informs the trainer.

### M136 AT4

**STANDARDS**: The firer performs the following actions:

### TRAINING AND EVALUATION

1. Ensures the launcher has no cracks, dents, or bulges, or reports these defects to the instructor.

2. Ensures the transport safety pin is in place, or reports this defect to the instructor.

3. Ensures the cocking lever is in the SAFE (uncocked) position, or reports this defect to the instructor.

4. Ensures the muzzle cover is intact, or reports this defect to the instructor.

5. Ensures the color-coded band is correct, or reports this defect to the instructor.

6. Ensures the sights function properly, or reports this defect to the instructor.

7. Ensures the venturi has no foreign objects, or reports this defect to the instructor.

8. Ensures the forward safety functions properly, or reports this defect to the instructor.

9. Ensures the red trigger button is not broken or missing, or reports this defect to the instructor.

10. Ensures the shoulder stop is not broken or damaged, or reports this defect to the instructor.

11. Ensures the carrying sling is not frayed and is firmly attached to the launch tube, or reports this defect to the instructor.

**RESULTS** **GO NO-GO**

Critique and grade the soldier.

**TASK 2**: Place a light antiarmor weapon into operation.

**CONDITIONS**: Given an expended light antiarmor weapon (FHT) on a range or in garrison.

**M72-Series LAW**

**STANDARDS**: The firer performs the following actions, in sequence:

**TRAINING AND EVALUATION**

1. Removes the rear cover pull pin.

2. Releases the rear cover and sling assembly.

3. Extends the launcher vigorously and ensures it is fully extended.

4. Places the launcher on his firing shoulder.

5. Checks the backblast area.

6. Pulls the arming handle out.

7. Sights.

**M136 AT4**

**STANDARDS**: The firer performs the following actions, in sequence:

**TRAINING AND EVALUATION**

1. Cradles the AT4 in his arms.

2. Removes the transport safety pin.

3. Unsnaps and unfolds the shoulder stop.

4. Places the AT4 on his shoulder.

5. Releases the sights.

6. Checks the backblast area.

7. Places cocking lever in the ARMED (cocked) position.

8. Fully depresses and holds down the forward safety.

**RESULTS** **GO** **NO-GO**

Critique and grade the soldier.

**TASK 3**: Demonstrate the standing position used to fire a light antiarmor weapon.

**CONDITIONS**: Given a target and an expended light antiarmor weapon in any suitable location.

**STANDARDS**: The firer demonstrates the standing position as follows:

## TRAINING AND EVALUATION

### M72-Series LAW

1. Faces the target by executing a half-left face turn and spreading feet a comfortable distance apart.

2. Places the launcher on the shoulder with the left hand directly under the forward portion of the launcher.

3. Places the right elbow against the body for stability.

### M136 AT4

1. Faces the target by executing a half-left face turn and spreads the feet a comfortable distance apart.

2. Places the launcher on the shoulder and grasps the sling near the launcher's muzzle with the left hand.

3. Places the right elbow against the body for stability.

| RESULTS | GO | NO-GO |
|---|---|---|
| Critique and grade the soldier. | | |

**TASK 4**: Demonstrate the kneeling position used to fire a light antiarmor weapon.

**CONDITIONS**: Given a target and an expended light antiarmor weapon in a suitable location, with the soldier in the standing position.

**STANDARDS**: The firer demonstrates the kneeling position as follows:

## TRAINING AND EVALUATION

1. Kneels on the right knee with the right thigh at a 90-degree angle to the line of aim.

2. Sits back on the right heel and shifts weight forward.

3. Rests the upper left arm forward of the left knee.

4. Places the right arm against the body.

**RESULTS** **GO NO-GO**

Critique and grade the soldier.

**TASK 5**: Demonstrate the sitting position used to fire a light antiarmor weapon.

**CONDITIONS**: Given a light antiarmor weapon training device or expended launcher in a suitable location, with the soldier in the standing position.

**STANDARDS**: Each firer demonstrates the sitting position as follows:

**TRAINING AND EVALUATION**

1. Keeps the launcher pointed toward the target.

2. Executes a half-left face turn and sits with the legs crossed.

3. Places the launcher on the right shoulder.

4. Leans slightly forward from the hips.

5. Rests the elbows forward of the knees to avoid bone-to-bone contact.

**RESULTS** **GO NO-GO**

Critique and grade the soldier.

**TASK 6**: Demonstrate the prone position used to fire a light antiarmor weapon.

**CONDITIONS**: Given a target and an expended light antiarmor weapon (FHT) in a suitable location.

**STANDARDS**: The firer demonstrates the prone position as follows:

**TRAINING AND EVALUATION**

1. To keep clear of the backblast area, lies down at an angle of not less than 45 degrees from the line of fire.

2. Keeps back straight and right leg directly on line with the right hip and right shoulder.

3. Moves left leg as far as possible to ensure comfort.

4. Holds both elbows well below the launcher.

5. Checks the backblast area to ensure that no part of the body is in it.

6. Holds head as steady as possible, with the firing eye lined up with the sights.

| | RESULTS | GO | NO-GO |
|---|---|---|---|

Critique and grade the soldier.

**TASK 7**: Identify enemy vehicles and weapons.

**CONDITIONS**: Given ten vehicle identification cards (GTA 17-2-13).

**STANDARDS**: The soldier correctly identifies eight out of ten vehicles or weapons by nomenclature.

| | RESULTS | GO | NO-GO |
|---|---|---|---|

Critique and grade the soldier.

**TASK 8**: Estimate range.

**CONDITIONS**: Given stationary, partly exposed, or fully exposed targets at ranges between 50 and 250 meters, during daylight, in weather conditions where all targets are visible. Targets include personnel, equipment, silhouettes, and vehicles.

**STANDARDS**: The soldier states the distance to each target with no more than 20 percent error.

| | RESULTS | GO | NO-GO |
|---|---|---|---|

Critique and grade the soldier.

**TASK 9**: Apply the correct lead for a target.

**CONDITIONS**: Given a sight template, a series of ten flank, frontal, and oblique silhouettes of armored targets, rates of movement, and ranges. (At least three different views, speeds, and ranges are given.)

**STANDARDS**: The soldier uses correct sighting procedures and correctly applies the lead eight out of ten times.

**RESULTS** **GO NO-GO**

Critique and grade the soldier.

**TASK 10**: Perform misfire procedures on a light antiarmor weapon.

**CONDITIONS**: Given a previously fired light antiarmor weapon and instructions that the weapon has misfired, on a suitable firing range, under simulated combat conditions, in any lighting conditions.

**M72-series LAW**

**STANDARDS**: The soldier applies misfire procedures as follows and tries to fire the M72-series LAW. If the weapon still fails to fire, he disposes of it IAW unit SOP:

**TRAINING AND EVALUATION**

1. Immediately squeezes the trigger bar again.

2. If the LAW still does not fire, tries to place the trigger arming handle on SAFE.

3. Removes the launcher from the shoulder and partly collapses it (10 to 15 centimeters or 4 to 6 inches), keeping the launcher pointed upward and downrange.

4. Extends the launcher.

5. If the LAW fails to fire again, squeezes the trigger, tries to return the weapon to SAFE, partly collapses the launcher, and sets the weapon aside.

**M136 AT4**

**STANDARDS**: The soldier performs combat misfire procedures on the AT4 as follows:

**TRAINING AND EVALUATION**

1. Releases the forward safety with the right hand.

2. Recocks the firing mechanism using the cocking lever.

3. Presses and holds down the forward safety, aims, and presses the trigger.

4. If launcher still fails to fire, repeats Steps 1 and 2.

5. If the launcher again fails to fire, returns the cocking lever to the SAFE (uncocked) position.

6. Removes the weapon from the shoulder, keeping it pointed toward the target.

7. Replaces the transport safety pin.

8. Informs the instructor that the firer breaks the sights off to indicate a misfired weapon when in combat.

9. Places the launcher on the ground, pointing toward the target.

| RESULTS | GO | NO-GO |
|---|---|---|

Critique and grade the soldier.

**TASK 11**: Return the light antiarmor weapon to its carrying configuration.

**CONDITIONS**: Given an expended light antiarmor weapon in a suitable location.

**STANDARDS**: The soldier performs the following actions, in sequence:

### M72-series LAW

### TRAINING AND EVALUATION

1. Pushes in the trigger arming handle.

2. Removes the launcher from the shoulder.

3. Depresses the barrel detent and collapses the launch tube, guiding the front and rear sights into position.

4. Replaces the sling assembly.

5. Replaces the rear cover pull pin.

### M136 AT4

1. Releases the forward safety.

2. Returns the cocking lever to the SAFE (uncocked) position by pushing forward and upward on it, then letting it spring back into position.

3. Removes the launcher from the shoulder, ensuring the muzzle is pointed in the direction of fire.

4. Replaces the transport safety pin, fully seating it in the retainer hole.

5. When closing the rear sight, resets the range indicator to the 200-meter setting. (If this step is not accomplished, the rear sight breaks off when the sight cover is closed.)

6. Restores the sights by laying them down and closing their covers.

7. Snaps the shoulder stop into the closed position.

8. Moves to another location; slings the launcher over either shoulder and moves out.

| | RESULTS | GO | NO-GO |
|---|---|---|---|

Critique and grade the soldier.

# APPENDIX D
# SUGGESTED DESIGN FOR COMBINED LIGHT ANTIARMOR RANGE

*This appendix provides basic guidance for units to establish a range complex for light antiarmor weapons. The complex described may be used for multilevel training and firing of light antiarmor weapons; both subcaliber trainers and live HE ammunition may be used on the same range. Training areas should be near, but not adjacent to, the firing line. The firing line should be designed to allow personnel to observe firing from the side. The target array should include stationary targets composed of F-type targets or the hulls of hard targets (APCs or tanks), a moving target (a target on track or a manned moving target tank [MMTT]), and bunkers. This allows the firer to engage a variety of targets and to develop his skills in the various types of engagements.*

## D-1. FIRING LINE

The firing line should be organized to allow for the 100-meter AT4 backblast (Figure D-1). Also, at least 20 meters must be allowed between firing points. The backblast area must be fenced, roped, or marked in some way, so soldiers know not to enter it when firing is being conducted. A range that meets the requirements of the AT4 will also meet the requirements of the LAW.

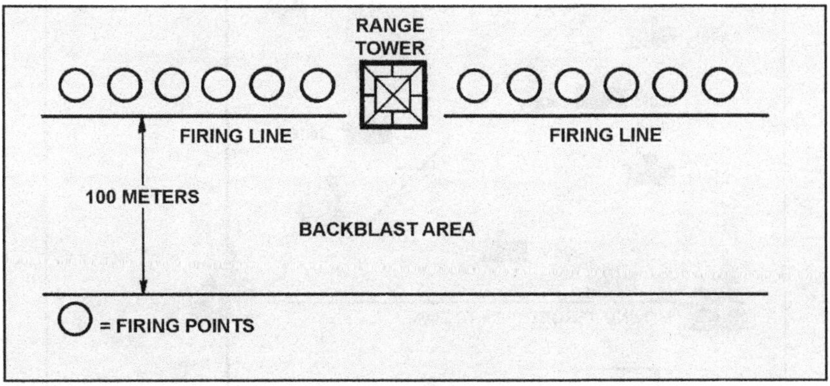


**Figure D-1. Firing line.**

## D-2. TARGET ARRAY

The target array should include targets at ranges of 100 to 700 meters (Figure D-2). This enables the unit to conduct advanced instructional fire on targets located beyond the weapons' maximum effective ranges. Standard vehicle silhouettes or tank hulls should be used for stationary targets. Selected targets should be positioned behind mounds of earth to simulate vehicles in defilade or partly behind trees or other objects. Firers adjust their points of aim accordingly. Targets should also be positioned to allow firers to engage flank, frontal, and oblique targets. Bunker targets should be placed within the target array at ranges of 100 to 400 meters. Bunker targets should be no more than 1 meter high or 2 1/2 to 3 meters wide. Moving target silhouettes should travel along a track or road so the firers can engage fast moving targets from both flank and oblique angles. The moving targets should be placed at ranges of 150 to 600 meters and should enter and exit from the flanks outside the range fan. Firers may only use HE weapons to engage hard targets. This reduces the damage to other targets the unit must maintain such as F-type silhouettes and bunkers.

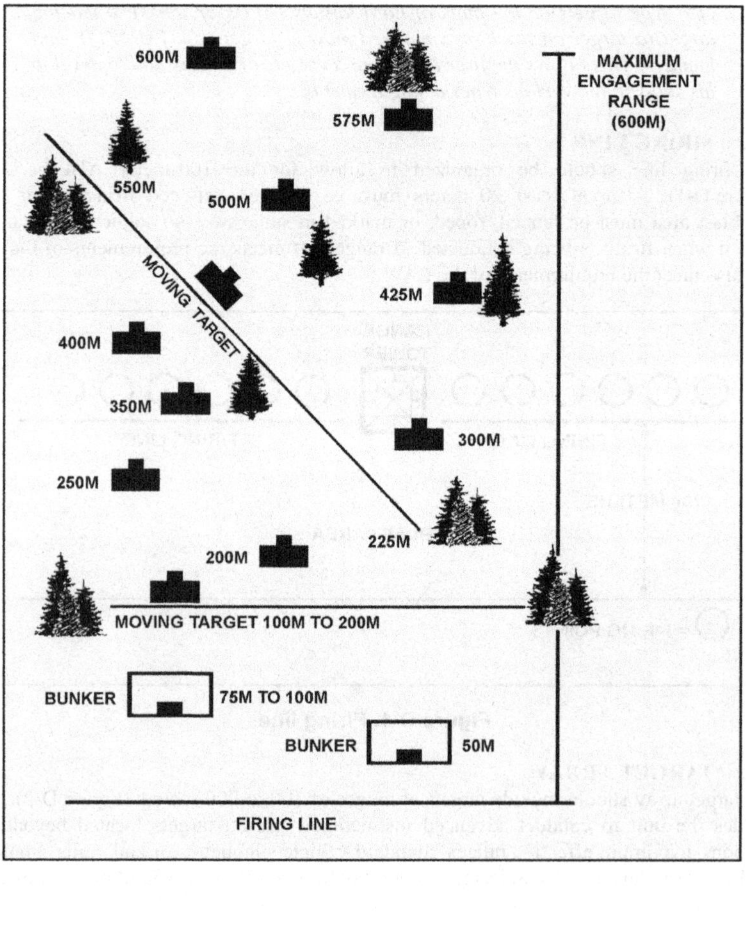

Figure D-2. Target array.

APPENDIX E

# FIRING TABLES AND EXAMPLE SCORECARDS

*This appendix provides firing tables for the subcaliber gunnery trainers discussed in this manual. It also includes an example of a completed scorecard for each light antiarmor weapon. All instructional and qualification firing tables are conducted on suitable ranges with ammunition designated in the firing tables. Some tables are fired in daylight, others in limited visibility, but all have some NBC firing (Table E-1).*

| | |
|---|---|
| **M72-Series LAW** | Instructional day fire, stationary targets |
| | Instructional day fire, moving targets |
| | Instructional night fire |
| **M136 AT4** | Instructional fire |
| | Instructional night fire |
| | Qualification fire |
| | Qualification night fire |
| | Advanced instructional fire |

**Table E-1. Firing tables for light antiarmor weapons.**

## E-1. M72-SERIES LAW

The M190 35-mm subcaliber device is used to fire the instructional firing tables for the M72-series LAW and DA Form 7323-R (M72-Series LAW Scorecard) is used to score them. Figure E-1 shows an example of the completed form. The instructional day and night firing tables for the LAW are provided in Tables E-2 and E-3.

**NOTE:** A blank copy of the form is located at the end of this publication for local reproduction on 8 1/2-inch by 11-inch paper.

## M72-SERIES LAW SCORECARD

Instructions for using this form are provided in FM 23-25.
TRADOC is the proponent agency.

NAME THIELE, STEPHEN SSN 123-45-6789 DATE 19 APR 94

GRADE E-5 RATING 1ST CLASS UNIT HHC 3/29 INF. REGT. WEATHER (circle one) CLEAR (RAIN) FOG SNOW

### DATA REQUIRED BY THE PRIVACY ACT OF 1974

**AUTHORITY:** 10 USC 3012(g) / Executive Order 9397 **PRINCIPAL PURPOSE(S):** Record individual performance. **ROUTINE USES:** Evaluate individual proficiency and determine proficiency level. SSN is used for positive identification purposes only. **MANDATORY OR VOLUNTARY DISCLOSURE AND EFFECT ON INDIVIDUAL NOT PROVIDING INFORMATION:** Voluntary. Individuals not providing information cannot be rated/scored on a mass basis.

| TABLE | TYPE OF FIRE | TYPE OF TARGET | ROUND | RANGE (M) | FIRING POSITION | HIT | MISS |
|---|---|---|---|---|---|---|---|
| 1 | Instructional DAY | Stationary | 1 | 125 to 150 | Standing | ✓ | |
| | | | 2 | 150 to 175 | Kneeling | ✓ | |
| | | | * 3 | 150 to 200 | Sitting | ✓ | |
| | | | * 4 | 125 to 150 | Prone | ✓ | |
| | | | | | Table Subtotals | 4 | 0 |

FIRER'S SIGNATURE DATE 19 APR 94 SCORER'S SIGNATURE DATE 19 APR 94

| TABLE | TYPE OF FIRE | TYPE OF TARGET | ROUND | RANGE (M) | FIRING POSITION | HIT | MISS |
|---|---|---|---|---|---|---|---|
| 2 | Instructional DAY | Moving | 1 | 100 to 200 | Standing | ✓ | |
| | | | 2 | 100 to 200 | Sitting | ✓ | |
| | | | * 3 | 100 to 200 | Kneeling | | ✓ |
| | | | | | Table Subtotals | 2 | 1 |

FIRER'S SIGNATURE DATE 20 APR 94 SCORER'S SIGNATURE DATE 20 APR 94

| TABLE | TYPE OF FIRE | TYPE OF TARGET | ROUND | RANGE (M) | FIRING POSITION | HIT | MISS |
|---|---|---|---|---|---|---|---|
| 3 | Instructional NIGHT | Stationary | 1 | 100 to 150 | Kneeling | | ✓ |
| | | | 2 | 100 to 150 | Sitting | ✓ | |
| | | | * 3 | 100 to 150 | Sitting | ✓ | |
| | | | | | Table Subtotals | 2 | 1 |

FIRER'S SIGNATURE DATE 21 APR 94 SCORER'S SIGNATURE DATE 21 APR 94

* Firer must wear a protective mask when firing this round.

### CONSOLIDATED SCORES

FIRER'S SIGNATURE DATE 21 APR 94 TOTAL MISSES 2

SCORER'S SIGNATURE DATE 21 APR 94 RANK SFC

RANGE OIC'S SIGNATURE DATE 21 APR 94 RANK CPT

| INSTRUCTOR NOTES | |
|---|---|
| ALL FIRING TABLES | • Do not announce the ranges to targets.<br>• Fire all engagements in sequence. |

DA Form 7323, JUL 94

**Figure E-1. Example completed DA Form 7323-R.**

| TABLE 1<br>M72-Series LAW Instructional Day Fire,<br>Stationary Targets | |
|---|---|
| **TASK** | Each soldier engages stationary armored targets with the M190 subcaliber device. |
| **CONDITIONS** | Given one M190 subcaliber device and four M73 rockets, on a suitable firing range with stationary targets at a range of 100 to 225 meters. Two targets are engaged in MOPP4. |
| **STANDARD** | The soldier fires four rockets at stationary targets and achieves at least three hits. |
| **TABLE 2<br>M72-Series LAW Instructional Day Fire,<br>Moving Targets** | |
| **TASK** | Each soldier engages moving armored targets with the M190 subcaliber device. |
| **CONDITIONS** | Given one M190 subcaliber device, three M73 rockets, and targets moving at a rate of 8 to 24 kmph and at a range of 100 to 200 meters. Frontal and oblique-moving targets may be used when a manned, moving tank is available. One target is engaged in MOPP4. |
| **STANDARD** | The soldier fires three rockets at moving targets and achieves at least two hits. |

**Table E-2. Instructional day firing tables for the M72-series LAW.**

| TABLE 3<br>M72-Series LAW Instructional Night Firing<br><br>Before you can use the AN/PVS-4, you must install an M72A1 sight reticle on it and Boresight it to an expended M72-series LAW launcher. | |
|---|---|
| **TASK** | Each soldier engages stationary targets at night with the M190 subcaliber device. |
| **CONDITIONS** | Given one M190 subcaliber device and three M73 rockets, on a suitable firing range with stationary targets at a range of 100 to 150 meters, at night, and either an AN/PVS-4 or illumination provided by indirect fire. |
| **STANDARD** | The soldier fires three rockets at stationary targets and achieves at least one hit. He engages one of these targets in MOPP4. |

**Table E-3. Instructional night firing table for the M72-series LAW.**

## E-2. M136 AT4

The M287 9-mm tracer bullet trainer is used to fire the instructional, qualification, and advanced instructional firing tables for the M136 AT4 and DA Form 7324-R (M136 AT4 Scorecard) is used to score them. A copy of this form is located at the end of this publication for local reproduction on 8 1/2-inch by 11-inch paper. Figure E-2 shows an example of the completed form; Tables E-4 through E-6, on pages E-7, E-8, and E-9, provide the instructional, qualification, and advanced instructional firing tables for the AT4. The latter must be fired on a separate, more specialized range.

## E-3. NIGHT COURSE OF FIRE TABLES AND QUALIFICATION FOR M136 AT4

Each soldier is given a PMI before the instructional and qualification tables are fired. Once the soldier completes Table I and III (day tables), he is given additional instruction on the use of the aided vision devices for the night operations.

a. **Qualification Tables for the M287 Tracer Trainer.**

- Table I (Instructional Day). This scenario consists of 12 targets from 100 to 300 meters. The standard to proceed to the next table is 6 out of 12 targets successfully engaged.
- Table II (Instructional Night). This scenario consists of 8 targets from 100 to 300 meters. The standard to proceed to the next table is 6 out of 12 targets successfully engaged at night.
- Table III (Qualification Day). This scenario consists of 8 targets from 100 to 300 meters. The standard to proceed to the next table is 6 out of 8 targets successfully engaged.
- Table IV (Qualification Night). This scenario consists of 8 targets from 100 to 300 meters (same as Table III, but conducted at night). The standard to proceed to the advanced firing tables is 6 out of 8 targets successfully engaged.

**NOTES:** 1. In Table IV, eight engagements are proposed as opposed to the current six engagements.
2. All references to the M136 launcher and cartridge refer to the M287 9-mm tracer bullet training device.

b. **Ammunition Requirements for Night Training Strategy.** The current STRAC ammunition allocation and course of fire is compared with the proposed night training strategy in Table E-7 on page E-9. The current STRAC table stipulates that 120 rounds of 9-mm tracer ammunition is authorized to each squad for additional squad members to fire Tables I and IV.

c. **Day Standards.** Commanders are able to achieve the current day standard at night with the M136 AT4 through the use of training strategy within this field manual. A gunner who is qualified on the day standard and has executed preliminary marksmanship training with the appropriate night vision device can achieve the current day standard at night. During testing, gunners had no problem engaging targets out to 300 meters. Table E-8, on page E-9, shows the day standard and is also the proposed night standard.

**M136 AT4 SCORECARD**

Instructions for using this form are provided in FM 23-25.
TRADOC is the proponent agency.

NAME DINKENS, DAVID SSN 135-24-5678 DATE 19 APR 94

GRADE E-3 RATING ______ UNIT HHC 2/29 INF REGT. WEATHER (circle one) CLEAR (RAIN) FOG SNOW

**DATA REQUIRED BY THE PRIVACY ACT OF 1974**

**AUTHORITY:** 10 USC 3012(g) / Executive Order 9397. **PRINCIPAL PURPOSE(S):** Record individual performance. **ROUTINE USES:** Evaluate individual proficiency and determine proficiency level. SSN is used for positive identification purposes only. **MANDATORY OR VOLUNTARY DISCLOSURE AND EFFECT ON INDIVIDUAL NOT PROVIDING INFORMATION:** Voluntary. Individuals not providing information cannot be rated/scored on a mass basis.

| TABLE | TYPE OF FIRE | TYPE OF TARGET | ROUND | RANGE (M) | FIRING POSITION | HIT | MISS |
|---|---|---|---|---|---|---|---|
| 1 | Instructional Day | Stationary | 1 | 150 to 200 | Kneeling | ✓ | |
| | | Stationary | * 2 | 150 to 200 | Kneeling | ✓ | |
| | | Stationary | 3 | 250 to 300 | Sitting | ✓ | |
| | | Stationary | 4 | 250 to 300 | Sitting | ✓ | |
| | | Stationary | 5 | 125 to 150 | Prone | ✓ | |
| | | Stationary | 6 | 125 to 150 | Prone | ✓ | |
| | | MMTT/MAS | 7 | 150 to 200 | Kneeling | ✓ | |
| | | MMTT/MAS | * 8 | 150 to 200 | Kneeling | ✓ | |
| | | MMTT/MAS | 9 | 200 to 250 | Standing | | ✓ |
| | | MMTT/MAS | *10 | 200 to 250 | Fighting position, unsupported | | ✓ |
| | | MMTT/MAS | 11 | 250 to 300 | Kneeling | ✓ | |
| | | MMTT/MAS | 12 | 250 to 300 | Kneeling | | ✓ |
| | | | | | Table Subtotals | 9 | 3 |

FIRER'S SIGNATURE David Dinkens DATE 19 APR 94 SCORER'S SIGNATURE Dan Sell DATE 19 Apr 94

| TABLE | TYPE OF FIRE | TYPE OF TARGET | ROUND | RANGE (M) | FIRING POSITION | HIT | MISS |
|---|---|---|---|---|---|---|---|
| 2 | Instructional Night | Stationary | 1 | 125 to 150 | Prone | | ✓ |
| | | Stationary | 2 | 125 to 150 | Prone | ✓ | |
| | | Stationary | * 3 | 150 to 200 | Sitting | ✓ | |
| | | Stationary | 4 | 150 to 200 | Sitting | ✓ | |
| | | Stationary | 5 | 250 to 300 | Kneeling | | ✓ |
| | | Stationary | 6 | 250 to 300 | Kneeling | ✓ | |
| | | MMTT/MAS | 7 | 125 to 150 | Standing | ✓ | |
| | | MMTT/MAS | 8 | 125 to 150 | Standing | ✓ | |
| | | MMTT/MAS | * 9 | 200 to 250 | Kneeling | ✓ | |
| | | MMTT/MAS | 10 | 200 to 250 | Kneeling | ✓ | |
| | | MMTT/MAS | 11 | 150 to 200 | Standing | | ✓ |
| | | MMTT/MAS | 12 | 150 to 200 | Fighting position, unsupported | ✓ | |
| | | | | | Table Subtotals | 9 | 3 |

FIRER'S SIGNATURE David Dinkens DATE 19 APR 94 SCORER'S SIGNATURE Dan Sell DATE 19 Apr 94

* Firer must wear a protective mask when firing this round.

| INSTRUCTOR NOTES | |
|---|---|
| ALL FIRING TABLES | • Do not announce the ranges to targets.<br>• Fire all engagements in sequence.<br>• Fire Table 1 before Table 3; fire Table 2 before Table 4. |
| FIRING TABLES 2 THROUGH 5 | • Ensure no one fires at front or rear of a manned moving tank target.<br>• Total hits and misses, and enter results in the consolidated scores block on the back of this scorecard. |

DA Form 7324-R, JUL 94 (FRONT)

**Figure E-2. Example completed DA Form 7324-R (front).**

**M136 AT4 SCORECARD**

| TABLE | TYPE OF FIRE | TYPE OF TARGET | ROUND | RANGE (M) | FIRING POSITION | HIT | MISS |
|---|---|---|---|---|---|---|---|
| | | Stationary | 1 | 125 to 150 | Standing | ✓ | |
| | | Stationary | 2 | 150 to 200 | Kneeling | ✓ | |
| | | Stationary | * 3 | 200 to 250 | Sitting | ✓ | |
| 3 | Qualification Day | Stationary | 4 | 250 to 300 | Sitting | ✓ | |
| | | MMTT/MAS | 5 | 125 to 150 | Kneeling | ✓ | |
| | | MMTT/MAS | 6 | 150 to 200 | Fighting position, unsupported | | ✓ |
| | | MMTT/MAS | * 7 | 200 to 250 | Sitting | ✓ | |
| | | MMTT/MAS | 8 | 250 to 300 | Kneeling | ✓ | |
| | | | | | Table Subtotals | 7 | 1 |

FIRER'S SIGNATURE *David Dinkens* DATE *21 Apr 94* SCORER'S SIGNATURE *Dan Sell* DATE *21 Apr 94*

| TABLE | TYPE OF FIRE | TYPE OF TARGET | ROUND | RANGE (M) | FIRING POSITION | HIT | MISS |
|---|---|---|---|---|---|---|---|
| | | Stationary | * 1 | 125 to 150 | Standing or Prone | ✓ | |
| | | Stationary | 2 | 200 to 250 | Kneeling | ✓ | |
| 4 | Qualification Night | Stationary | 3 | 250 to 300 | Sitting | ✓ | |
| | | MMTT/MAS | 4 | 150 to 200 | Sitting | ✓ | |
| | | MMTT/MAS | * 5 | 125 to 150 | Kneeling | | ✓ |
| | | MMTT/MAS | 6 | 200 to 250 | Fighting position, unsupported | ✓ | |
| | | | | | Table Subtotals | 5 | 1 |

FIRER'S SIGNATURE *David Dinkens* DATE *21 APR 94* SCORER'S SIGNATURE *Dan Sell* DATE *21 Apr 94*

| TABLE | TYPE OF FIRE | TYPE OF TARGET | ROUND | RANGE (M) | FIRING POSITION | HIT | MISS |
|---|---|---|---|---|---|---|---|
| | | Sequence | 1 | | Sitting | ✓ | |
| | | Sequence | 2 | | Sitting | ✓ | |
| 5 | Advanced Instructional | Pair | 3 | | Kneeling | ✓ | |
| | | Pair | 4 | | Kneeling | ✓ | |
| | | Volley | 5 | | Kneeling | ✓ | |
| | | Volley | * 6 | | Sitting | ✓ | |
| | | | | | Table Subtotals | 6 | 0 |

FIRER'S SIGNATURE *David Dinkens* DATE *22 Apr 94* SCORER'S SIGNATURE *Dan Sell* DATE *22 Apr 94*

* Firer must wear a protective mask when firing this round.

QUALIFICATION CRITERIA — BASED ON TOTAL HITS FOR TABLES 3 AND 4

FIRST CLASS ......... 13 HITS
SECOND CLASS ......... 11 HITS (circled)
THIRD CLASS ......... 9 HITS
UNQUALIFIED ......... FEWER THAN 9 HITS

**C O N S O L I D A T E D   S C O R E S**

INSTRUCTIONAL TOTALS (TABLES 1 AND 2): HITS *18* MISSES *6* CLASS RATING, TABLES 1 AND 2: *FIRST*

QUALIFICATION TOTALS (TABLES 3 AND 4): HITS *12* MISSES *2* OVERALL TOTALS HITS *30* MISSES *8*

FIRER'S SIGNATURE *David Dinkens* DATE *22 Apr 94* SCORER'S SIGNATURE *Dan Sell* DATE *22 Apr 94*

RANGE OIC'S SIGNATURE *[signature]* RANK *SFC* DATE *22 Apr 94*

DA Form 7324-R, JUL 94 (BACK)

**Figure E-2. Example completed DA Form 7324-R (back).**

| | |
|---|---|
| **TABLE 1**<br>**M136 AT4 Instructional Fire** | |
| **TASK** | Each soldier engages an armored target with the M287 tracer bullet trainer. |
| **CONDITIONS** | Given one M287 9-mm subcaliber device and twelve rounds of 9-mm tracer ammunition, a suitable firing range, stationary targets at a range of 100 to 300 meters, and other targets moving 8 to 24 kmph at a range of 150 to 200 meters. When a manned, moving tank is available, frontal and oblique-moving targets may be used. |
| **STANDARD** | The soldier fires twelve tracer bullets at stationary and moving targets. He engages three of these targets in MOPP4. |
| **TABLE 3**<br>**M136 AT4 Instructional Fire** | |
| **TASK** | Each soldier engages an armored target with the M287 tracer bullet trainer at night. |
| **CONDITIONS** | At night, given one M287 9-mm subcaliber device and twelve rounds of 9-mm tracer ammunition, artificial illumination, a suitable firing range, stationary targets at a range of 100 to 300 meters, and other targets moving from 8 to 24 kmph (5 to 15 mph) at a range of 150 to 200 meters. When a manned, moving tank is available, frontal and oblique-moving targets may be used. |
| **STANDARD** | The soldier fires twelve tracer bullets at stationary and moving targets. He engages two of these targets in MOPP4. |

**Table E-4. Instructional firing tables for the M136 AT4.**

| TABLE 2<br>M136 AT4 Qualification Fire | |
|---|---|
| **TASK** | Each soldier engages an armored target with the M287 tracer bullet trainer. |
| **CONDITIONS** | Given one M287 9-mm subcaliber device and eight rounds of 9-mm tracer ammunition, a suitable firing range, stationary targets at a range of 100 to 300 meters, and other targets moving at a rate of 8 to 24 kmph (5 to 15 mph) at a range of 150 to 200 meters. When a manned, moving tank is available, frontal and oblique-moving targets may be used. |
| **STANDARD** | The soldier fires eight tracer bullets at stationary and moving targets and achieves at least six hits. He engages two of these targets in MOPP4. |
| **TABLE 4<br>M136 AT4 Qualification Fire** | |
| **TASK** | Each soldier engages an armored target with the M287 tracer bullet trainer. |
| **CONDITIONS** | At night, given one M287 9-mm subcaliber device and six rounds of 9-mm tracer ammunition, artificial illumination, a suitable firing range, stationary targets at a range of 100 to 300 meters, and other targets moving 8 to 24 kmph (5 to 15 mph) at a range of 150 to 200 meters. When a manned, moving tank is available, frontal and oblique-moving targets may be used. |
| **STANDARD** | At night, given one M287 9-mm subcaliber device and six rounds of 9-mm tracer ammunition, artificial illumination, a suitable firing range, stationary targets at a range of 100 to 300 meters, and other targets moving 8 to 24 kmph (5 to 15 mph) at a range of 150 to 200 meters. When a manned, moving tank is available, frontal and oblique-moving targets may be used. |

**Table E-5. Qualification firing tables for the M136 AT4.**

| TABLE 5<br>M136 AT4 Advanced Instructional Fire<br>Fire this table on an appropriate separate, specialized range. | |
|---|---|
| **TASK** | Each soldier engages an armored target with the M287 tracer bullet trainer. |
| **CONDITIONS** | Given one M287 9-mm subcaliber device and six rounds of 9-mm tracer ammunition, a suitable firing range, stationary targets at a range of 100 to 300 meters, and other targets moving 8 to 24 kmph (5 to 15 mph) at a range of 150 to 200 meters. When a manned, moving tank is available, frontal and oblique-moving targets may be used. |
| **STANDARD** | The soldier must fire six tracer bullets at stationary and moving targets and achieve at least three hits. He must engage two of the targets with sequence firing, two with pair firing, and two with volley firing, and he must observe MOPP4 while engaging one of the six. |

**Table E-6. Advanced instructional firing table for the M136 AT4.**

| EVENT | CURRENT STRAC | RECOMMENDED CHANGES | TOTAL |
|---|---|---|---|
| TABLE I | 12 | 0 | 12 |
| TABLE II | 12 | 0 | 12 |
| TABLE III (QUALIFICATION) | 8 | 0 | 8 |
| TABLE IV (NIGHT QUALIFICATION) | 6 | 2 | 8 |
| TABLE V (ADVANCE) | 6 | 0 | 0 |
| TOTAL | 44 | 2 | 40 |

**Table E-7. Ammunition requirements for night training strategy.**

| M136 AT4 (M287 TRAINER) TABLE III/IV | | | |
|---|---|---|---|
| **ENGAGEMENT** | **RANGE** | **POSITION** | **HIT/MISS** |
| 1 | 125 to 150 | Standing | |
| 2 | 150 to 200 | Kneeling | |
| 3 | 200 to 250 | Sitting | |
| 4 | 250 to 300 | Sitting | |
| 5 | 125 to 150 | Kneeling | |
| 6 | 150 to 200 | Fighting Position | |
| 7 | 200 to 250 | Sitting | |
| 8 | 250 to 300 | Kneeling | |

**Table E-8. M136 AT4 (M287 trainer) Table III/IV.**

APPENDIX F

# INFRARED LASER AIMING DEVICES

*The newest infrared laser pointers increase the night firing accuracy of infantry weapons. The new family of weapon mounts (some fielded with the thermal weapon sight) and helmet brackets are more stable and reliable for continued accurate firing under field conditions. Their effectiveness is limited by the capability of the image-intensifying ($I^2$) sight with which they are used. Normally this is a night vision goggle or monocular, with or without the 3X magnifier. Again, they primarily enhance the gunner's ability to align his sights on the target, but usually have dramatic affect on his firing positions and scanning technique. Unlike $I^2$ sights mounted on the weapon, the gunner can somewhat reduce the adverse effects of muzzle flash by rising up higher off the sight with his helmet-mounted goggle or monocular. With the AN/PEQ-2A, the range of these devices is greatly improved, allowing great effectiveness for crew-served weapons out to 2,200 meters. However, trainers and leaders must adopt new safety procedures to ensure these lasers, which are not safe for the eyes, are treated like loaded weapons any time the batteries are issued. Finally, the bore light is also effective to boresight these pointers. Generally weapons do not require zeroing with live ammunition to confirm alignment after soldiers and leaders are qualified in the boresighting techniques. The AT4 cannot be conventionally boresighed, therefore a sight alignment method is used to zero the aided vision device.*

## F-1. AN/PAQ-4C, AIMING LIGHT

The AN/PAQ-4C aiming light (Figure F-1) projects an infrared laser beam, which cannot be seen with the eye but can be seen with night vision devices. This aiming light works with the AN/PVS-7B/C/D-series goggles and AN/PVS-14. The AN/PAQ-4C mounts on various weapons with mounting brackets and adapters.

a. **Technical Data**. The following technical data apply to the AN/PAQ-4C:

| | |
|---|---|
| Optics: | 100% parallax free, anti-reflective coated lens system |
| Length (sight): | 14 centimeters (5.5 inches) |
| Weight: | 164 grams (5.78 ounces) |
| Height: | 3 centimeters (1.2 inches) |
| Width: | 6.5 centimeters (2.5 inches) |
| Ranged: | Beyond 600 meters. (Actual range depends on light level and night vision device used for observation.) |
| Battery life: | 100-hour operating (ON) time for AA batteries in temperatures above 0 degrees Centigrade (32 degrees Fahrenheit); 36 hours for temperatures below 0 degrees Centigrade (32 degrees Fahrenheit) |

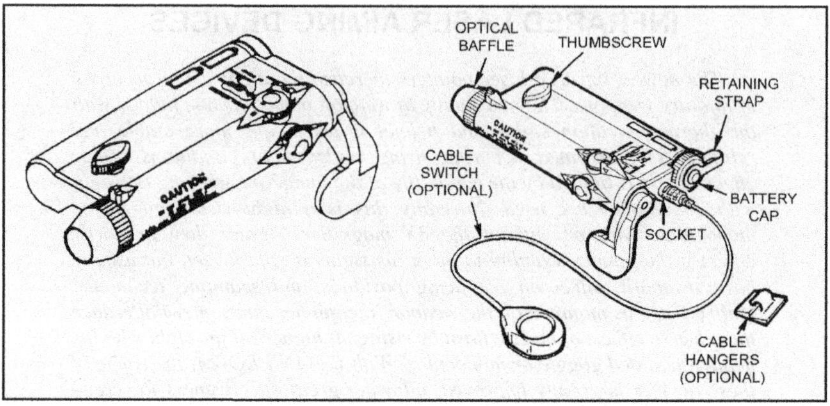


**Figure F-1. AN/PAQ-4C, aiming light.**

b. **Operation.** Activate the AN/PAQ-4C aiming light by rotating the ON/OFF switch lever or the button on the optional cable switch. Either switch connects power from two AA batteries to an internal electronic circuit, which produces the infrared laser. Internal lenses focus the infrared light into a narrow beam. Control the direction of the beam by rotating the mechanical adjusters with click detents. These adjusters are used to zero the aiming light to the weapon. Once zeroed to the weapon, the aiming light projects the beam along the line of fire of the weapon. The optical baffle prevents off-axis viewing of the aiming light beam by the enemy.

(1) ***M136 (AT4) Launcher and Cartridge Mounting Procedures.*** The aiming light is attached to the M136 (AT4) launcher and cartridge by first attaching the mounting bracket to the weapon and then attaching the aiming light to the mounting bracket (Figures F-2 and F3).

**STEP 1**: Install the mounting bracket.

**WARNING**

Only M136 (AT4) qualified personnel should install and use the mounting bracket assembly. Conduct all M136 (AT4) operator preventive maintenance checks and services before installation of mounting bracket assembly.

**NOTE:** All references to the M136 (AT4) launcher and cartridge are also applicable to the M287 9-mm tracer bullet training device.

(a) Cradle the M136 (AT4) in left arm.

(b) Position the support bracket with the mounting rail on the left side and the marking FRONT over the rear sight.

(c) With the pivot bracket spread open, place the support bracket against the base of the rear sight housing and the bottom on the shoulder strap boss.

(d) Swing the pivot bracket around the M136 (AT4) and secure it by rotating the locking latch clockwise to engage the latch shaft.

(e) The lever screw assembly must be located in the rear threaded screw hole when mounting the aiming light. Lever screw assembly may require relocation from the front to the rear threaded screw hole.

(f) Place the bracket adapter (Figure F-2) in groove of the mounting rail so that the threaded screw hole in the base of the adapter is aligned with the lever screw assembly and tighten the lever screw.

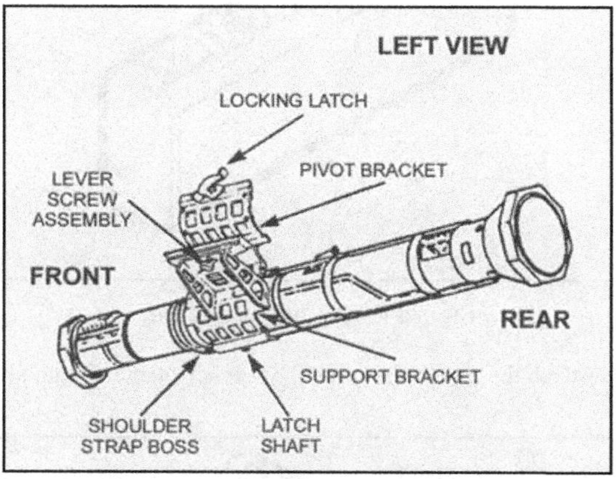


**Figure F-2. M136 (AT4) installation.**

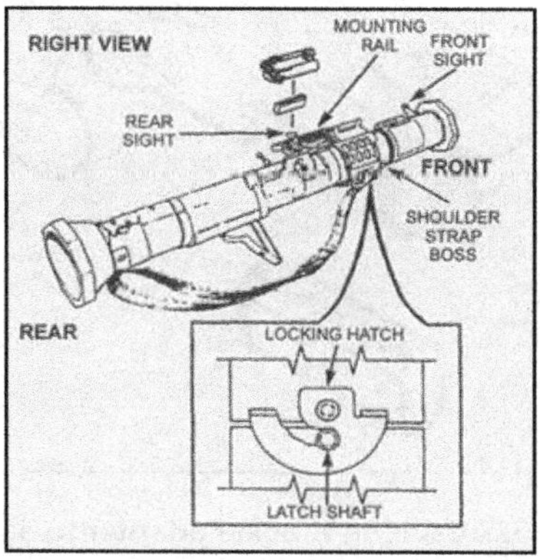


**Figure F-3. M136 (AT4) installation (locking latch and shaft).**

**STEP 2**: Install the bracket adapter (Figure F-4). The bracket adapter provides compatibility between the aiming light and the mounting brackets used on the M2, M60, and M136 AT4 weapons.

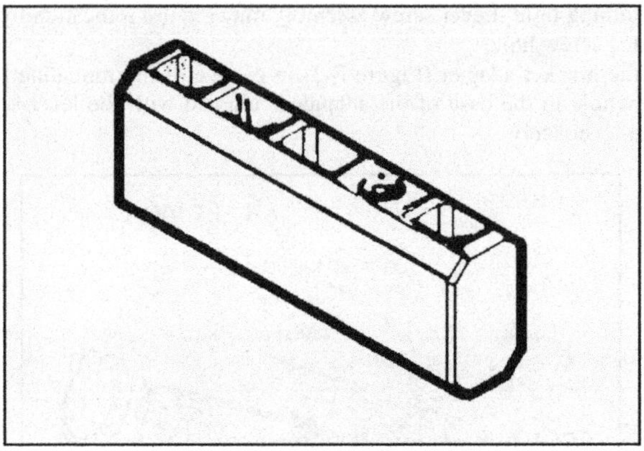

**Figure F-4. Bracket adapter.**

**STEP 3**: Attach the AN/PAQ-4 onto the bracket adapter (Figure F-5).

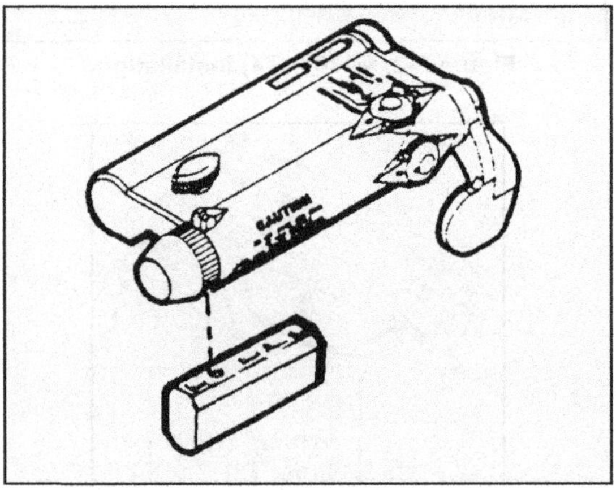

**Figure F-5. Attaching bracket adapter to aiming light.**

**STEP 4**: Attach the AN/PAQ-4 to the AN/PVS-4 mounting bracket.

- Position the AN/PAQ-4 on top of the bracket so that the spacer of the AN/PAQ-4 is aligned with the mounting knob of the bracket.
- Turn the mounting knob clockwise until the AN/PAQ-4 is tight.

**WARNING**
**Do not store the AN/PAQ-4 with batteries installed.**

(2) ***M136 (AT4) Launcher and Cartridge Dismounting Procedures.***

**STEP 1**: Detach the AN/PAQ-4 from the AN/PVS-4 mounting bracket.

- Turn the mounting knob counterclockwise until the AN/PAQ-4 is loose.
- Remove the AN/PAQ-4 from the bracket adapter.

**STEP 2**: Remove the adapter from the AN/PAQ-4

**STEP 3**: Remove the mounting bracket.

- Rotate the locking latch counter clockwise to disengage the latch shaft.
- Swing pivot bracket away from around the AT4.

## F-2. AN/PEQ-2A TARGET POINTER/ILLUMINATOR/AIMING LIGHT

The AN/PEQ-2A target pointer/illuminator/aiming light (TPIAL) (Figure F-6) is a Class IIIb laser that emits a collimated beam of infrared light for precise aiming of the weapon as well as a separate infrared illuminating beam with adjustable focus. A safety block is provided for training purposes (blue side), which limits the operator from selecting the high power modes (black side). The TPIAL projects an infrared laser beam that cannot be seen with the eye but can be seen with night vision devices. It is also capable of projecting a much wider infrared illuminating beam from an integral illuminator. The TPIAL works with night vision goggles and mounts on various weapons with mounting brackets and adapters. The AN/PEQ-2A can also be used in the hand-held mode to illuminate and designate targets by leaders.

a. **Technical Data.**

| | |
|---|---|
| Weight: | 7.5 ounce (with 2 AA batteries) |
| Length: | 16.26 centimeters (6.4 inches) |
| Width: | 7.12 centimeters (2.8 inches) |
| Height: | 3.05 centimeters (1.2 inches) |
| Range: | 600 meters in low power (eye safe) |
| | 2,000 meters in high power (non-eye safe) |
| Output Power: | |
| Aiming laser | 25 m W (+- 10%) |
| Illuminator | 30 m W (+50,-20%) |
| Beam Divergence: | |
| Aiming laser | .5 m rad |
| Illuminator | 1.0-115 m rad |
| Batteries: | 2 x 1.5V AA batteries |

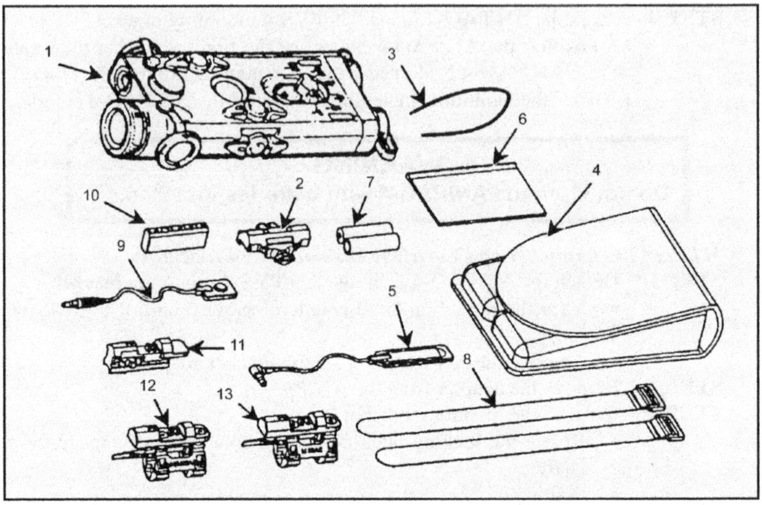


| | |
|---|---|
| (1) TPIAL ASSEMBLY WITH SAFETY BLOCK | (6) OPERATOR'S MANUAL |
| (2) RAIL | (7) BATTERIES 1.5V AA |
| (3) NECK CORD | (8) STRAP, RETENTION |
| (4) TEXTILE BAG | (9) CABLE SWITCH, 20-INCH, BUTTON |
| (5) CABLE SWITCH, 12-INCH, MEMBRANE | (10) BRACKET ADAPTER |
| | (11) TRAINING EXTENDER (ARMY ONLY) |
| | (12) M4/M16A2 BRACKET ASSEMBLY |

**Figure F-6. AN/PEQ-2A with accessories.**

b. **Operation.** This paragraph describes the battery installation, the safety block installation, the mode switch, the button switch, the cable switch, the focus switch, the lens cap, and the boresight adjusters.

(1) ***Battery Installation*** (Figure F-7).

(a) Unscrew the battery caps and install two AA batteries.

(b) Orient the batteries as indicated by the markings on the AN/PEQ-2A body.

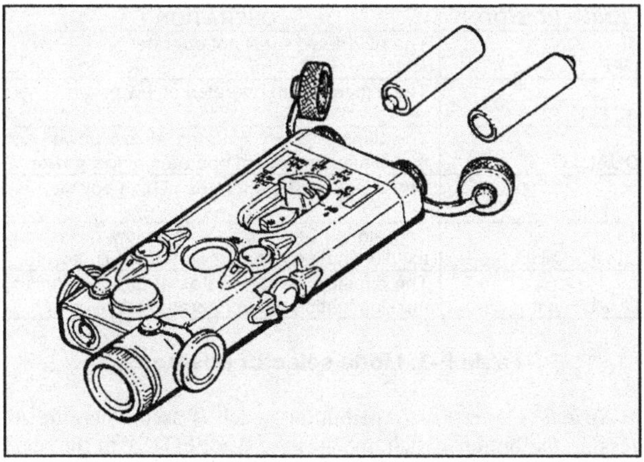

**Figure F-7. AN/PEQ-2A battery installation.**

(2) ***Safety Block Installation*** (Figure F-8). The safety block installed in the training mode (blue side up) prevents the operator from accessing the noneye safe modes (AIM HI, DUAL LO/HI, DUAL HI/HI). A .050 hex head Allen wrench is needed to unscrew the block from the body and reinstall it in the tactical mode (black side up).

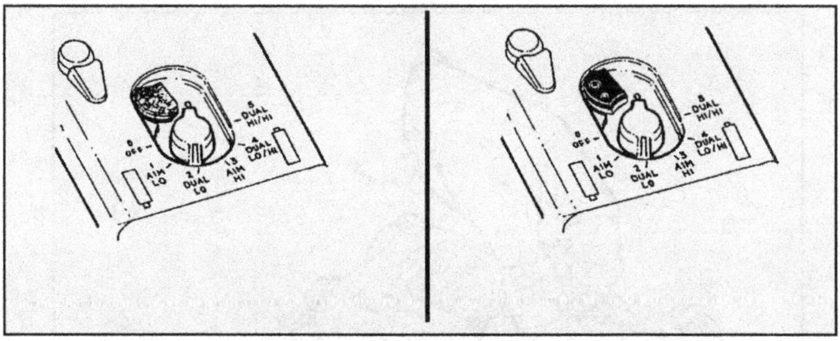

**Figure F-8. Safety block installation.**

(3) ***Mode Selector***. The mode selector is used to set the mode in which the AN/PEQ-2A operates when the cable switch button or push button are depressed. The mode selector has six positions (Table F-1).

| KNOB POSITION | OPERATION |
|---|---|
| 0<br>OFF | The AN/PEQ-2A will not operate. |
| 1<br>AIM LO | The aiming beam operates at low power. |
| 2<br>DUAL LO | The aiming beam operates at low power and the illuminating beam operates at low power. |
| 3<br>AIM HI | The aiming beam operates at high power. |
| 4<br>DUAL LO/HI | The aiming beam operates at low power and the illuminating beam operates at full power. |
| 5<br>DUAL HI/HI | The aiming beam operates at high power and the illuminating beam operates at full power. |

**Table F-1. Mode selector positions.**

(4) ***Button Switch*** (Figure F-9). The button switch is used when the AN/PEQ-2A is hand held. Pressing the button switch operates the AN/PEQ-2A in the operational mode set by the mode selector. When the button is released, the AN/PEQ-2A turns off.

(a) A green light emitting diode (LED) is incorporated into the body of the AN/PEQ-2A to indicate that the AN/PEQ-2A is ON. Whenever the AN/PEQ-2A is activated, the green LED will light and stay lit until the unit is turned OFF.

(b) If continuous operation of the AN/PEQ-2A is desired, pressing the button switch twice in rapid succession will latch the AN/PEQ-2A ON. The AN/PEQ-2A will remain on until the push button is pressed a third time.

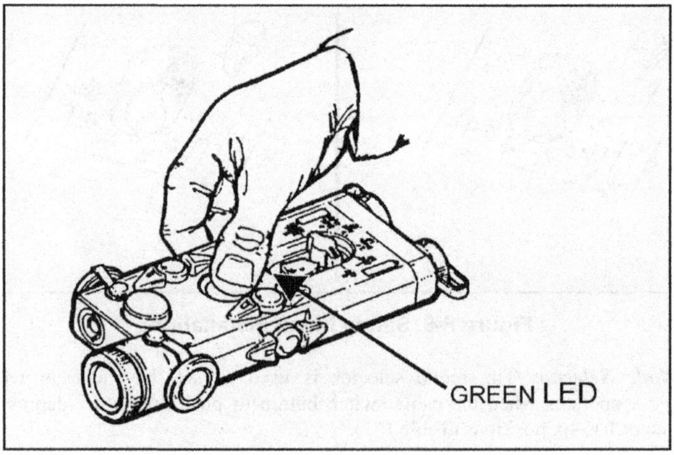


**Figure F-9. Operation of the button switch.**

(5) ***Cable Switch*** (Figure F-10). The cable switch is used when the AN/PEQ-2A is mounted on a weapon. The cable switch plugs into the back of the AN/PEQ-2A

assembly. Pressing the button or pad at the end of the cable switch causes the AN/PEQ-2A to turn on in the operational mode selected by the mode select switch. When the button is released, the AN/PEQ-2A turns off.

(a) If continuous operation of the AN/PEQ-2A is desired, pressing the cable switch twice in rapid succession will latch the AN/PEQ-2A ON. The AN/PEQ-2A will remain on until the push button is pressed a third time.

(b) When the cable switch plug is installed in the AN/PEQ-2A, it automatically locks into place. To remove the switch, pull back on the plug sleeve and pull the plug out. DO NOT TRY TO REMOVE THE PLUG BY PULLING ON THE CABLE.

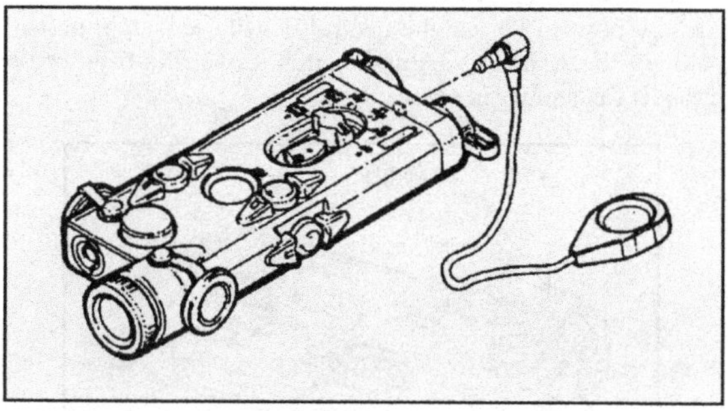

**Figure F-10. Installation of the cable switch.**

(6) ***Focus Knob*** (Figure F-11). The focus knob is used to vary the spread of the illumination beam based on the range and size of the area to be illuminated.

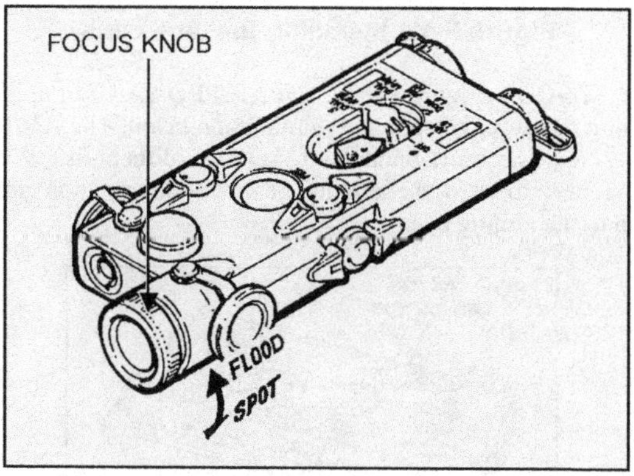


**Figure F-11. Using the focus knob.**

(7) ***Lens Cap*** (Figure F-12).

(a) The *black lens cap* blocks the AN/PEQ-2A illuminator or aiming laser beam should the device be activated. To use the black lens cap, pull it from its stored location on the side of the AN/PEQ-2A and stretch it over the front of the focus knob or aiming beam so that it fits snugly in place.

(b) The *diffuser lens cap* enables the illuminator or aiming laser to emit in a 45-degree cone (10 feet at 10 feet). To use the diffuser lens cap, pull it from its stored location on the side of the AN/PEQ-2A and stretch it over the front of the focus knob or aiming beam so that it fits snugly in place.

(c) The *neutral density lens cap* enables the AN/PEQ-2A illuminator or aiming laser to be operated in low power. To use the neutral density lens cap, pull it from its stored location on the side of the AN/PEQ-2A and stretch it over the front of the focus knob or aiming beam so that it fits snugly in place.

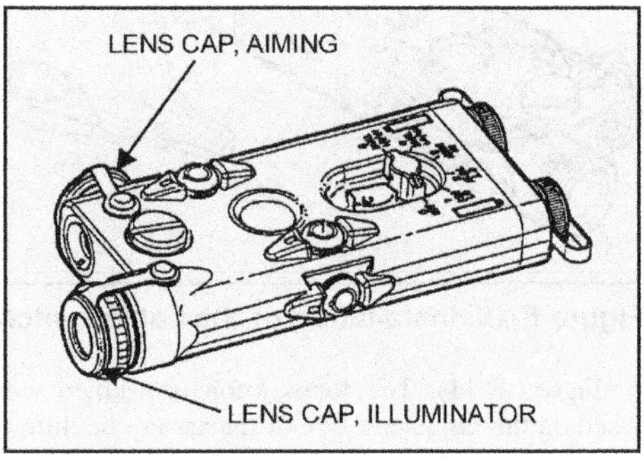


**Figure F-12. Installing the lens caps.**

(8) ***Boresight Adjusters*** (Figure F-13). The AN/PEQ-2A is equipped with boresight adjusters for zeroing the aiming beam and illumination beam. The AN/PEQ-2A adjusters move the beams in true horizontal and vertical directions. When zeroing the AN/PEQ-2A, it is best to zero the aiming beam to the weapon and then align the illumination beam to the aiming beam.

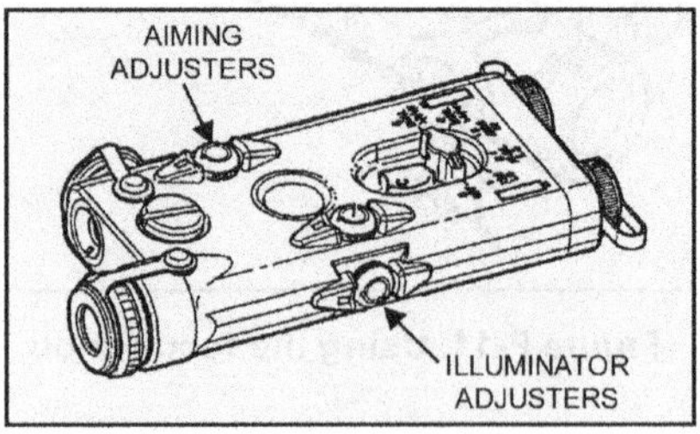


**Figure F-13. Boresight adjusters.**

**F-3. AN/PAS-13 (V2) MEDIUM WEAPON THERMAL SIGHT and AN/PAS-13 (V3) HEAVY WEAPON THERMAL SIGHT**

The AN/PAS-13 (V2) medium weapon thermal sight (MWTS) and the AN/PAS-13 (V3) heavy weapon thermal sight (HWTS) (Figure F-14) are silent, lightweight, compact, and durable battery-powered infrared imaging sensors that operate with low battery consumption.

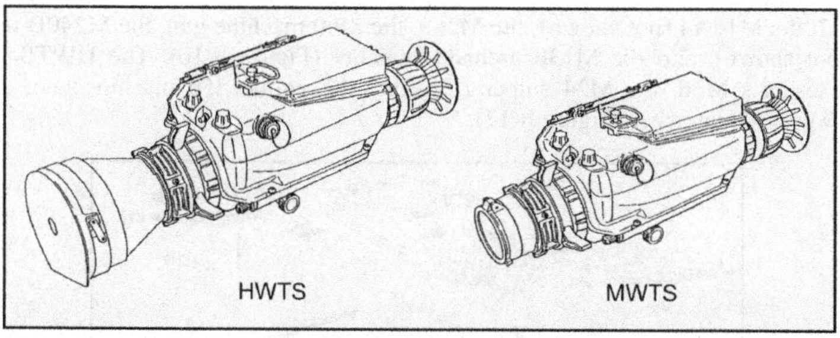


**Figure F-14. HWTS and MWTS models of the thermal weapon sight.**

a. **Description.**

(1) Both the MWTS and the HWTS are referred to as singular thermal weapon sights (TWS). The TWS is capable of target acquisition under conditions of limited visibility such as darkness, smoke, fog, dust, and haze. The TWS operates effectively at night and can also be used during the daytime. Infrared light is received through the telescope, detected by an IR sensor, converted to digital data, processed, and then displayed for the user. The TWS is composed of two functional groups: the telescope and the basic sensor.

(a) *Telescope.* The telescope receives IR light emitting from an intended target and its surroundings. The telescope magnifies and projects the IR light onto the scanner on the basic sensor (Figure F-15).

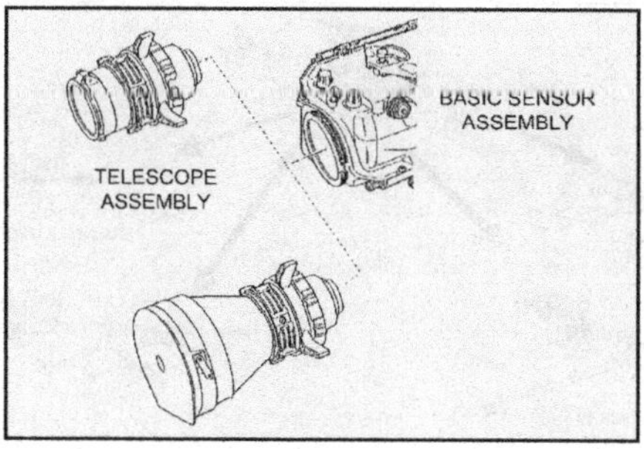


**Figure 3-6. TWS configurations.**

(b) *Basic sensor*. The scanner reflects the IR light received from the telescope onto the detective assembly. The detective assembly senses the IR light and converts it to video. The sensor's electronics condition the video for display on the LED array. The LED array illuminates the IR image along with the reticle. The light from the LED array is reflected off the scanner to form an image at the eyepiece.

(2) The only difference between the MWTS and the HWTS are the telescopes, which are different magnifications and hold different reticles. The basic sensor on the two models is the same. The MWTS fits the M4 carbine, the M4 modular weapon system, the M16A2, the M16A4 (not shown), the M249, the M60 machine gun, the M240B machine gun (not shown), and the M136 grenade launcher (Figure F-16). The HWTS fits the squad leader's M16, the M24 sniper rifle, the M2, caliber .50 machine gun, and the MK-19 grenade launcher (Figure F-17).

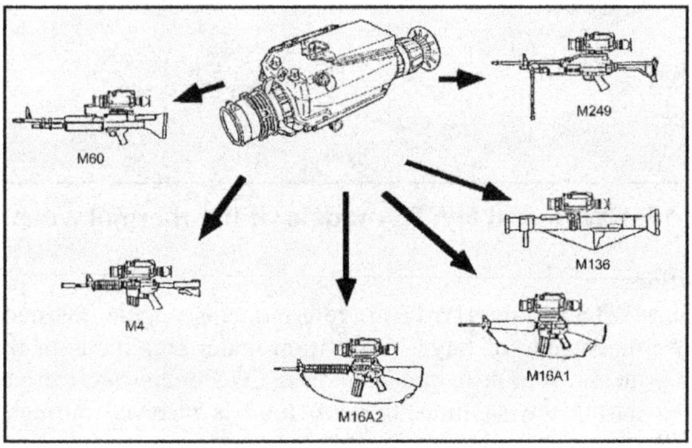


**Figure F-16. MTWS.**

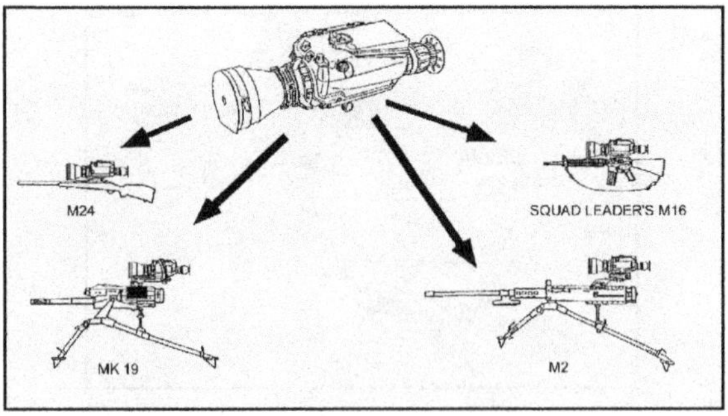


**Figure F-17. HTWS.**

b. **Technical Data.**

| | MWTS | HWTS |
|---|---|---|
| Field of view (FOV): | | |
| Narrow | 9 degrees | 3 degrees |
| Wide | 15 degrees | 9 degrees |
| Telescope magnification: | | |
| Narrow FOV | 3.3X | 10X |
| Wide FOV | 2X | 3.3X |
| Length: | 15.5 inches | 18 inches |
| Width: | 6.25 inches | 6.25 inches |
| Height: | 6.25 inches | 6.25 inches |
| Weight | 4.1 pounds | 4.5 pounds |
| Power consumption: | | |
| ON mode | 12.0 watts | 12.0 watts |
| STANDBY mode | 5.5 watts | 5.5 watts |
| Battery life: | | |
| 71% STANDBY mode | 10 hours | 10 hours |
| EMERGENCY mode | 3.5 hours | 3.5 hours |

c. **Operation.**

(1) The TWS has three modes of operation: STANDBY, ON, and EMERGENCY.

(a) STANDBY mode. When the system is first turned on, the TWS begins a cool down period of approximately two minutes. After the cool down period, the TWS enters the STANDBY mode. During the STANDBY mode, power is not applied to the scanner or display in order to extend the life of the battery.

(b) ON mode. When the TWS is in the STANDBY mode and pressure is applied to the eyecup, the TWS switches to the ON mode, and a switch engages to provide power to the scanner and display. After a three-second delay, the system is fully operational.

(c) EMERGENCY mode. When switched to the EMERGENCY mode, the TWS continuously applies power to the entire system. This allows the operator to bypass the three-second delay experienced when switching from the STANDBY to the ON mode. Since power is applied to the entire system while in the emergency mode, battery life is greatly reduced.

(2) Controls and indicators are as shown in Figure 3-18.

(a) The BRIGHTNESS CONTROL is a nine-position rotary switch with an off detent position (turned fully counterclockwise). The purpose is to turn the system on or off and adjust the brightness of the eyepiece display.

(b) The CONTRAST CONTROL adjusts the contrast of the thermal image displayed on the raster. It has an automatic and a manual mode.

(c) The FOCUS RING adjusts the telescope focus from 20 meters to infinity. It requires a manual adjustment and affects both the wide and narrow fields of view.

(d) The FIELD OF VIEW RING is located on the telescope. It has a wide and a narrow field of view. The wide FOV is for using low magnification during target detection, and the narrow FOV is for using high magnification during recognition and engagement.

(e) The RETICLE SELECT SWITCH selects one of the available reticles depending on the TWS model (medium or heavy). It must be held for two seconds to enable reticle changes..After two seconds, release the switch to cycle to the next reticle. This control is disabled after ten seconds of inactivity.

(f) The RETICLE ADJUST SWITCH adjusts the reticle aiming features in azimuth and elevation. It is used during zeroing, and it must be held for two seconds to allow changes to be made. After two seconds, each press moves the reticle aiming features one increment. This control is also disabled after ten seconds of inactivity.

(g) The BLACK/WHITE POLARITY SWITCH selects the polarity of the thermal image displayed on the raster. The initial setting is "white hot." The polarity switch affects the appearance of the target.

(h) The DIOPTER FOCUS RING adjusts the focus of the raster and indicators to the operator's eye. It ranges from +2 to -6 diopters.

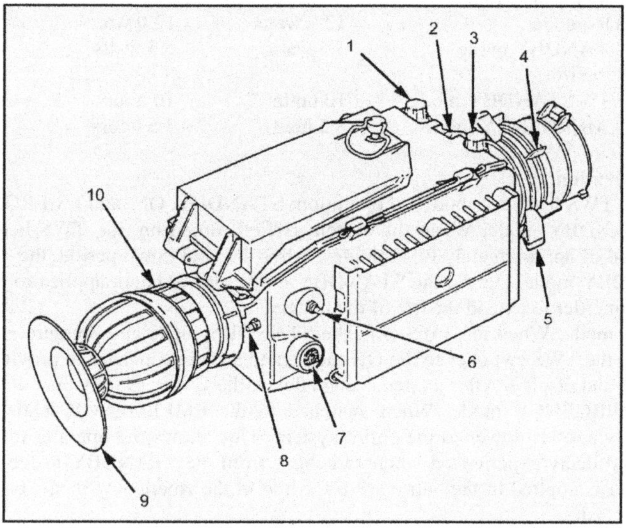


| | | | |
|---|---|---|---|
| (1) | BRIGHTNESS CONTROL | (6) | RETICLE SELECT SWITCH |
| (2) | EMERGENCY SWITCH CONTROL | (7) | RETICLE ADJUST SWITCH |
| (3) | CONTRAST CONTROL | (8) | BLACK/WHITE POLARITY SWITCH |
| (4) | FOCUS RING | (9) | EYECUP |
| (5) | FIELD OF VIEW RING | (10) | DIOPTER FOCUS RING |

**Figure F-18. TWS controls and indicators.**

(i) The EYEPIECE INDICATORS (Figure F-19) illuminate as follows:

- NOT COOL when the detectors are not cool enough for proper operation.
- WHT HOT/BLK HOT polarity.
- EMER during emergency mode operation.
- LOW when battery power has approximately fifteen minutes of useful power left.

(j) The COOLDOWN PERIOD INDICATORS blink the first ten seconds after the system is turned on. After ten seconds, the NOT COOL indicator is lit and the POLARITY, EMERGENCY, and LOW BATTERY indicators function normally.

(k) The DISPLAY RASTER displays the thermal image with a superimposed reticle.

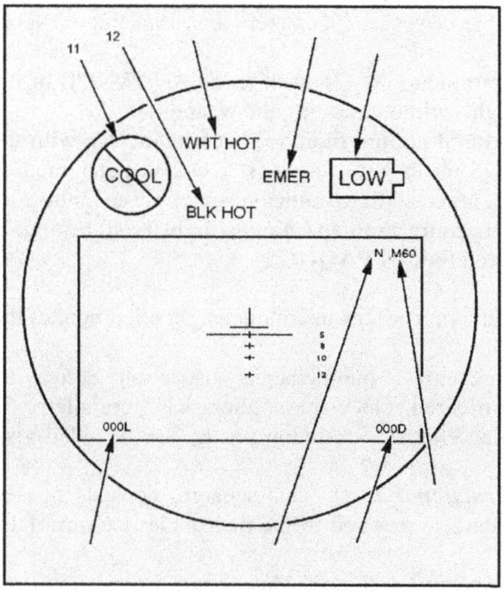


**Figure F-19. Eyepiece indicators.**

## F-4. M136 AT4/M287 AT4 SUBCALIBER TRACER TRAINER SIGHT ALIGNMENT PROCEDURES

The fundamentals of marksmanship are the same as during the daytime with two exceptions. When using aided vision devices, the firer should hold the mounting bracket with the left hand. This method provides greater stability and helps to balance the AT4 against the added weight of the sight. When aiming the AT4, remember to aim by placing the infrared aiming light center mass on the target.

> **WARNING**
> **When firing the M136 AT4, do not place your face within 3 inches of the mounted sight. The recoil of the AT4 could cause injury you or damage to the equipment.**

a. Select a stable position for the weapon.

b. Open the M136 front and rear sight covers. During limited visibility conditions, the 7-millimeter peephole may be necessary for zeroing.

c. Set the rear sight for 200 meters and choose the appropriate scenario.

(1) Place a 25 meter zero target with a predetermined offset at 10 meters and align the weapon's fixed sight to coincide with the 4-centimeter aiming point of the zero target.

(2) Select a suitable target at 200 meters and align the weapon's fixed sight center mass of the target.

d. Turn on the infrared aiming light. Use the AN/PVS-7B or AN/PVS-14 to observe the infrared aiming light (without moving the weapon).

e. Adjust the infrared aiming light so that it coincides with the 2-centimeter offset aiming point of the 25-meter zero target. If the 200-meter method is used, adjust the aiming point of the mounted sight to coincide center mass on the 200-meter target.

f. Repeat the procedure until the aiming light is sight aligned to the weapon by adjusting the controls on the AN/PAQ-4C.

**NOTE:** Use the same procedures to conduct sight alignment of the AN/PEQ-2.

g. Conduct a preventive maintenance inspection before the instructional and qualification tables are fired. Once the soldiers have completed Tables I and III (day tables), they receive additional instruction on the use of aided vision devices for night operations.

(1) ***Table I (Instructional Day)***. This scenario consists of 12 targets from 100 to 300 meters. The standard to proceed to the next table is 6 out of 12 targets successfully engaged.

(2) ***Table II (Instructional Night)***. This scenario consists of 12 targets from 100 to 300 meters. The standard to proceed to the next table is 6 out of 12 targets successfully engaged at night.

(3) ***Table III (Qualification Day)***. This scenario consists of 10 targets from 100 to 300 meters. The standard to proceed to the next table is 8 out of 10 targets successfully engaged.

(4) ***Table IV (Qualification Night)***. This scenario consists of 10 targets from 100 to 300 meters. The standard to proceed to the next table is 6 out of 10 targets successfully engaged.

h. The night sustainment training strategy should be conducted over a one-day time frame and consists of the following:

(1) ***Night Vision Goggles Familiarization Training***. Soldiers receive instruction on the proper use and fit of night vision goggles, to include characteristics and capabilities, maintenance, and mounting procedures.

(2) ***Infrared Aiming Light Familiarization Training***. Soldiers boresight the weapon and aided vision device at a range of 10 meters.

(3) ***Fundamentals of Firing (Body Position with Head Harness and NVD, Foxhole and Prone)***. Soldiers review and practice firing positions and fundamentals of marksmanship and any changes that may occur by use of the infrared aiming device.

(4) ***Weapon and Equipment Precombat Inspection***. The weapon and infrared aiming device is inspected and properly mounted. Inspect to ensure that the infrared aiming

device is mounted securely to the mount and that the mount is tightly secured to the weapon. A review of clearing or misfire procedures is important.

(5) ***Infrared Aiming Light Boresight to Weapon***. Each soldier boresights the weapon and infrared aiming device at a range of 10 meters. If a bore light is not available, then zero the weapon and sight by using the specified zeroing procedure for that weapon/sight combination.

# GLOSSARY

| | |
|---|---|
| AC | Active Component |
| ANCOC | advanced noncommissioned officer course |
| AN/PVS | Army Navy passive vision sight |
| ARTEP | Army Training and Evaluation Program |
| backblast | The backward blast created by igniting the propellant in a weapon designed to be as recoilless as possible. |
| blast | An explosive or violent detonation that produces a direct effect on its target in addition to radiant heat and overpressure at its point of origin. |
| BMP | a type of non-US fighting vehicle built by the former Soviet Union |
| BNCOC | basic noncommissioned officer course |
| BRDM | a type of non-US scout car built by the former Soviet Union |
| BTR | a type of non-US personnel carrier built by the former Soviet Union |
| C | Centigrade |
| CS | combat support |
| CSS | combat service support |
| DA | Department of the Army |
| DKIE | decontamination kit individual equipment |
| DOD | Department of Defense |
| DS | direct support |
| F | Fahrenheit |
| FHT | field handling trainer |

| | |
|---|---|
| fps | feet per second |
| FTX | field training exercise |
| HE | high-explosive |
| HEAT | high-explosive antitank |
| HELLFIRE | heliborne laser fire and forget (a missile system) |
| HMTS | medium weapon thermal sight |
| HWTS | heavy weapon thermal sight |
| IAW | in accordance with |
| IET | initial entry training |
| IR | infrared |
| kmph | kilometers per hour |
| LAW | light antitank weapon |
| LED | light emitting diode |
| METL | mission-essential task list |
| MILES | Multiple Integrated Laser Engagement System |
| mm | millimeter(s) |
| MMTT | manned moving target tank |
| MOPP | mission-oriented protective posture |
| MOUT | military operations on urbanized terrain |
| mph | miles per hour |
| mps | meters per second |
| MTP | mission training plan |

| | |
|---|---|
| MWTS | medium weapon thermal sight |
| NA | not applicable |
| NATO | North Atlantic Treaty Organization |
| NBC | nuclear, biological, chemical |
| NCO | noncommissioned officer |
| NSN | national stock number |
| NVD | night vision device |
| OIC | officer in charge |
| overpressure | increased atmospheric pressure (positive overpressure), followed by a wave of decreased atmospheric pressure (negative overpressure), produced around the origin of an explosive or violent detonation |
| PLDC | primary leadership development course |
| POI | program of instruction |
| promethium | a radioactive metallic element of the rare-earth group used to illuminate range markings on the front sight of the M72-series LAW |
| RC | Reserve Component |
| RPG | rocket-propelled grenade (non-US) |
| STX | situational training exercise |
| SMCT | soldier's manual of common tasks |
| SOP | standing operating procedure |
| spall | small fragment or chip |
| STP | soldier training publication |

| | |
|---|---|
| STRAC | standards in training commission |
| TOC | tactical operations center |
| TP | training-practice |
| TPIAL | target pointer/illuminator/aiming light, AN/PEQ-2A |
| TRADOC | (US Army) Training and Doctrine Command |
| trc | track (as used in STRAC manual) |
| TRP | target reference point |
| TSC | training support center |
| TWS | thermal weapon sight |

# REFERENCES

**SOURCES USED**
These are the sources quoted or paraphrased in this manual.


| | |
|---|---|
| TM 9-1055-886-12&P | Operator and Unit Maintenance Manual for M287 9-MM Tracer Bullet Training Device. 25 October 1989. |
| TM 9-1315-886-12 | Operator's and Unit Maintenance Manual for Launcher and Cartridge, 84 Millimeter: M136 (AT4). 15 May 1990. |
| TM 9-1340-214-10 | Operator's Manual for 66-MM Light Antitank Weapon (LAW) System M72A1, M72A2 with Coupler, M72A3, and Practice Rocket Launcher M190 with M73 Practice Rocket. 20 February 1992. |
| TM 11-5855-213-10 | Operator's Manual for Night Vision Sight, Individual-Served Weapon AN/PVS-4. 1 February 1993. |
| TM 11-5855-213-23&P | Unit and Direct Support Maintenance Manual for Night Vision Sight, Individual-Served Weapon, AN/PVS-4. 1 June 1993. |


**DOCUMENTS NEEDED**
These documents must be available to the intended users of this publication.


| | |
|---|---|
| DA Form 7323-R | M72-Series LAW Scorecard. July 1994. |
| DA Form 7324-R | M136 AT4 Scorecard. July 1994. |
| DA Pamphlet 350-38 | Standards in Weapons Training. 3 July 1997. |
| *DOD 5100.76-M | Physical Security of Sensitive Conventional Arms, Ammunition, and Explosives. 3 February 1983. |
| FM 7-7 | The Mechanized Infantry Platoon and Squad (APC). 15 March 1985. |
| FM 7-7J | The Mechanized Infantry Platoon and Squad (Bradley). 7 May 1993. |
| FM 7-8 | The Infantry Platoon and Squad. 1 March 2001. |
| FM 25-101 | Battle Focused Training. 30 September 1990. |

## M72-SERIES LAW SCORECARD

For use of this form, see FM 23-25; the proponent agency is TRADOC

NAME ______ SSN ______ DATE ______

GRADE ______ RATING ______ UNIT ______ WEATHER (circle one) CLEAR RAIN FOG SNOW

### DATA REQUIRED BY THE PRIVACY ACT OF 1974

**AUTHORITY:** 10 USC 3012(g) / Executive Order 9397 **PRINCIPAL PURPOSE(S):** Record individual performance. **ROUTINE USES:** Evaluate individual proficiency and determine proficiency level. SSN is used for positive identification purposes only. **MANDATORY OR VOLUNTARY DISCLOSURE AND EFFECT ON INDIVIDUAL NOT PROVIDING INFORMATION:** Voluntary. Individuals not providing information cannot be rated/scored on a mass basis.

| TABLE | TYPE OF FIRE | TYPE OF TARGET | ROUND | RANGE (M) | FIRING POSITION | HIT | MISS |
|---|---|---|---|---|---|---|---|
| 1 | Instructional DAY | Stationary | 1 | 125 to 150 | Standing | ___ | ___ |
| | | | 2 | 150 to 175 | Kneeling | ___ | ___ |
| | | | * 3 | 150 to 200 | Sitting | ___ | ___ |
| | | | * 4 | 125 to 150 | Prone | ___ | ___ |
| | | | | | Table Subtotals | ___ | ___ |

FIRER'S SIGNATURE ______ DATE ______ SCORER'S SIGNATURE ______ DATE ______

| TABLE | TYPE OF FIRE | TYPE OF TARGET | ROUND | RANGE (M) | FIRING POSITION | HIT | MISS |
|---|---|---|---|---|---|---|---|
| 2 | Instructional DAY | Moving | 1 | 100 to 200 | Standing | ___ | ___ |
| | | | 2 | 100 to 200 | Sitting | ___ | ___ |
| | | | * 3 | 100 to 200 | Kneeling | ___ | ___ |
| | | | | | Table Subtotals | ___ | ___ |

FIRER'S SIGNATURE ______ DATE ______ SCORER'S SIGNATURE ______ DATE ______

| TABLE | TYPE OF FIRE | TYPE OF TARGET | ROUND | RANGE (M) | FIRING POSITION | HIT | MISS |
|---|---|---|---|---|---|---|---|
| 3 | Instructional NIGHT | Stationary | 1 | 100 to 150 | Kneeling | ___ | ___ |
| | | | 2 | 100 to 150 | Sitting | ___ | ___ |
| | | | * 3 | 100 to 150 | Sitting | ___ | ___ |
| | | | | | Table Subtotals | ___ | ___ |

FIRER'S SIGNATURE ______ DATE ______ SCORER'S SIGNATURE ______ DATE ______

* Firer must wear a protective mask when firing this round.

### CONSOLIDATED SCORES

FIRER'S SIGNATURE ______ DATE ______ TOTAL MISSES ______

SCORER'S SIGNATURE ______ DATE ______ RANK ______

RANGE OIC'S SIGNATURE ______ DATE ______ RANK ______

| INSTRUCTOR NOTES | |
|---|---|
| ALL FIRING TABLES | • Do not announce the ranges to targets.<br>• Fire all engagements in sequence. |

DA Form 7323-R, JUL 94

**M136 AT4 SCORECARD**

For use of this form, see FM 23-25; the proponent agency is TRADOC

NAME ______________________ SSN ______________________ DATE ____________

GRADE ______ RATING ______ UNIT ______________ WEATHER (circle one) CLEAR RAIN FOG SNOW

**DATA REQUIRED BY THE PRIVACY ACT OF 1974**

**AUTHORITY:** 10 USC 3012(g) / Executive Order 9397. **PRINCIPAL PURPOSE(S):** Record individual performance. **ROUTINE USES:** Evaluate individual proficiency and determine proficiency level. SSN is used for positive identification purposes only. **MANDATORY OR VOLUNTARY DISCLOSURE AND EFFECT ON INDIVIDUAL NOT PROVIDING INFORMATION:** Voluntary. Individuals not providing information cannot be rated/scored on a mass basis.

| TABLE | TYPE OF FIRE | TYPE OF TARGET | ROUND | RANGE (M) | FIRING POSITION | HIT | MISS |
|---|---|---|---|---|---|---|---|
| 1 | Instructional Day | Stationary | 1 | 150 to 200 | Kneeling | ___ | ___ |
| | | Stationary | * 2 | 150 to 200 | Kneeling | ___ | ___ |
| | | Stationary | 3 | 250 to 300 | Sitting | ___ | ___ |
| | | Stationary | 4 | 250 to 300 | Sitting | ___ | ___ |
| | | Stationary | 5 | 125 to 150 | Prone | ___ | ___ |
| | | Stationary | 6 | 125 to 150 | Prone | ___ | ___ |
| | | MMTT/MAS | 7 | 150 to 200 | Kneeling | ___ | ___ |
| | | MMTT/MAS | * 8 | 150 to 200 | Kneeling | ___ | ___ |
| | | MMTT/MAS | 9 | 200 to 250 | Standing | ___ | ___ |
| | | MMTT/MAS | *10 | 200 to 250 | Fighting position, unsupported | ___ | ___ |
| | | MMTT/MAS | 11 | 250 to 300 | Kneeling | ___ | ___ |
| | | MMTT/MAS | 12 | 250 to 300 | Kneeling | ___ | ___ |
| | | | | | Table Subtotals | ___ | ___ |

FIRER'S SIGNATURE ______________________ DATE ________ SCORER'S SIGNATURE ______________________ DATE ________

| TABLE | TYPE OF FIRE | TYPE OF TARGET | ROUND | RANGE (M) | FIRING POSITION | HIT | MISS |
|---|---|---|---|---|---|---|---|
| 2 | Instructional Night | Stationary | 1 | 125 to 150 | Prone | ___ | ___ |
| | | Stationary | 2 | 125 to 150 | Prone | ___ | ___ |
| | | Stationary | * 3 | 150 to 200 | Sitting | ___ | ___ |
| | | Stationary | 4 | 150 to 200 | Sitting | ___ | ___ |
| | | Stationary | 5 | 250 to 300 | Kneeling | ___ | ___ |
| | | Stationary | 6 | 250 to 300 | Kneeling | ___ | ___ |
| | | MMTT/MAS | 7 | 125 to 150 | Standing | ___ | ___ |
| | | MMTT/MAS | 8 | 125 to 150 | Standing | ___ | ___ |
| | | MMTT/MAS | * 9 | 200 to 250 | Kneeling | ___ | ___ |
| | | MMTT/MAS | 10 | 200 to 250 | Kneeling | ___ | ___ |
| | | MMTT/MAS | 11 | 150 to 200 | Standing | ___ | ___ |
| | | MMTT/MAS | 12 | 150 to 200 | Fighting position, unsupported | ___ | ___ |
| | | | | | Table Subtotals | ___ | ___ |

FIRER'S SIGNATURE ______________________ DATE ________ SCORER'S SIGNATURE ______________________ DATE ________

* Firer must wear a protective mask when firing this round.

| INSTRUCTOR NOTES | |
|---|---|
| **ALL FIRING TABLES** | • Do not announce the ranges to targets.<br>• Fire all engagements in sequence.<br>• Fire Table 1 before Table 3; fire Table 2 before Table 4. |
| **FIRING TABLES 2 THROUGH 5** | • Ensure no one fires at front or rear of a manned moving tank target.<br>• Total hits and misses, and enter results in the consolidated scores block on the back of this scorecard. |

# INDEX

FIELD MANUAL
NO. 3-23.24

*FM 3-23.24 (FM 23-24)
HEADQUARTERS
DEPARTMENT OF THE ARMY
WASHINGTON, DC, 30 August 2001

# M47 DRAGON
# MEDIUM ANTITANK WEAPON SYSTEM

## CONTENTS



**DISTRIBUTION RESTRICTION:** Approved for public release; distribution is unlimited.

* This publication supersedes FM 23-24, 3 April 1990.

# PREFACE

This manual provides guidance for using and employing the M47 Dragon, a medium-range, antiarmor, surface-attack, guided missile weapon system. It provides information about the M47 Dragon's characteristics, nomenclature, and functions. The training program provided helps develop, evaluate, and maintain the proficiency of Dragon gunners. Tactical employment procedures described herein do not cover all situations.

Leaders and members of units equipped with the M47 Dragon should use this manual to train and test individual gunners or teams. This manual also discusses five key elements for maintaining combat proficiency with the Dragon:

- Technical and tactical information.
- A training program that stresses command involvement.
- Centralized training.
- Gunner selection.
- Development of qualified trainers.

The proponent of this publication is the United States Army Infantry School. Send comments and recommendations on DA Form 2028 (Recommended Changes to Publications and Blank Forms) directly to the Commandant, US Army Infantry School, ATTN: ATSH-INB-O, Fort Benning, Georgia 31905-5594 or to lusanoh@benning.army.mil.

Unless otherwise stated, whenever the masculine gender is used, both men and women are included.

CHAPTER 1
# INTRODUCTION

*This chapter describes the M47 Dragon and its components and provides pertinent technical data. The Dragon is a surface-attack, wire-guided, man-portable, shoulder-fired, medium antitank weapon system. It can defeat armored vehicles, fortified bunkers, concrete gun emplacements, and other hardened targets. The individual soldier or a two-man team can operate this weapon. Mechanized infantry gunners can use the Dragon with the M175 guided missile launcher mount. This mount provides a stable platform on either the M113 APC or on the M3 or M122 machine gun tripod. The Dragon can engage targets in daylight (with the daysight) and in limited visibility conditions, such as smoke, fog, or darkness (with the nightsight).*

## 1-1. MAJOR COMPONENTS

The Dragon weapon system consists of a daysight, a nightsight, and a round of ammunition.

a. **Daysight**. The daysight is an electro-optical sight that determines the position of the missile relative to the gunner's line of sight (LOS).

(1) The daysight generates signals that travel by wire to the missile control system. The missile control system fires the appropriate rocket thrusters to keep the missile heading along the gunner's LOS (Figure 1-1).

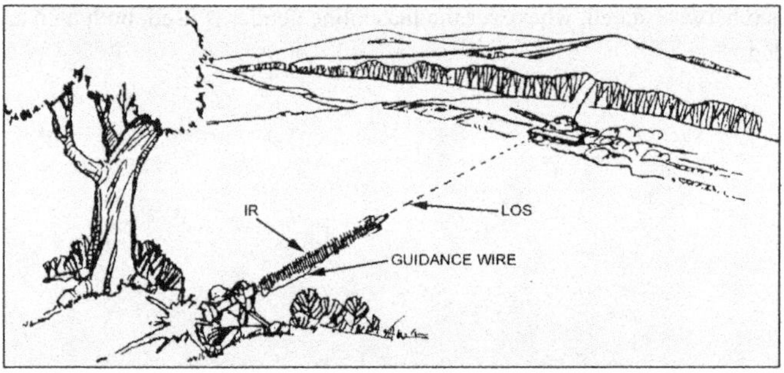


**Figure 1-1. Daysight line of sight.**

(2) The daysight attaches to and removes quickly from the round (Figure 1-2, page 1-2). The daysight's aluminum housing assembly contains the lens, infrared receiver assembly, and control signal comparator. The firing mechanism, located on the right side of the daysight's housing, consists of a trigger safety plunger and trigger lever bar. A rubber boot seals out moisture from the firing mechanism and safety plunger.

(3) A connector cover provides both mechanical and moisture protection for the electrical connector. A lanyard attaches the cover to the daysight.

(4) Shock absorbers surround the forward and rear ends of the daysight to protect it and its lens. The lens cover is attached to the daysight by a lanyard.

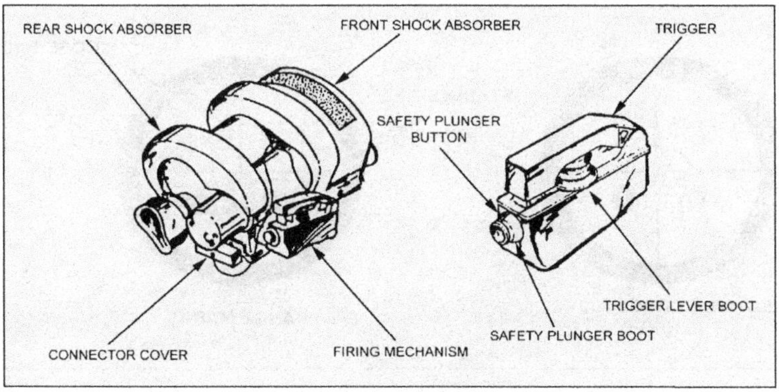


**Figure 1-2. Daysight parts.**

(5) The daysight's lens is a 6-power telescope. It magnifies the image across a 6-degree field of view (FOV) to help the gunner locate, identify, and track targets. A fitted and molded rubber eye guard protects the gunner's eyes and allows him to adjust rapidly to the sight (Figure 1-3). The infrared receiver consists of an objective lens assembly, graded filter, infrared detector, and nutator mirror. The sensor detects the missile's infrared output while discriminating against background signals. It detects any deviations or excursions of the missile from the gunner's LOS.

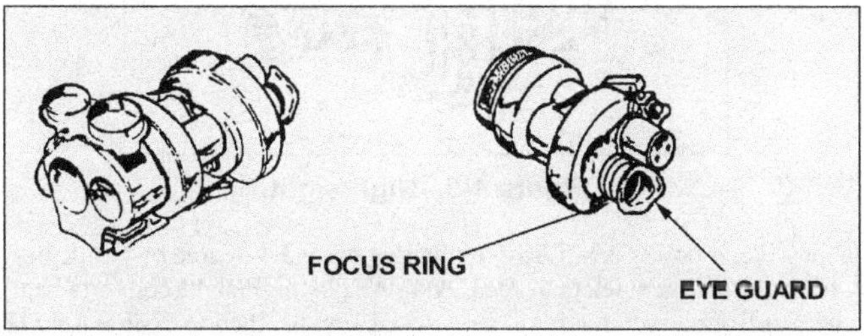


**Figure 1-3. Daysight lens and rubber eye guard.**

(6) The daysight's telescopic lens and infrared receiver align to define the LOS to the target. The tapered cross hair helps the gunner concentrate on the center of his FOV. Then, he can quickly place the cross hairs on the target's center of mass. He uses the stadia lines to determine the range to the target (Figure 1-4). For example, the stadia lines show how large a standard 3-meter by 6-meter Soviet-built vehicle would appear at the Dragon's maximum engagement range of 1,000 meters.

b. **Nightsight**. At dusk, or anytime the light level drops below what is needed to use the daysight, Dragon gunners switch to the nightsight (Figure 1-5). It attaches to the round the same as the daysight.

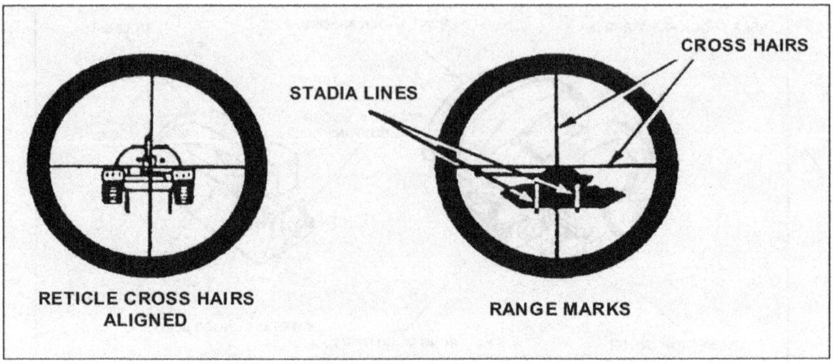


**Figure 1-4. Daysight reticle and stadia lines.**

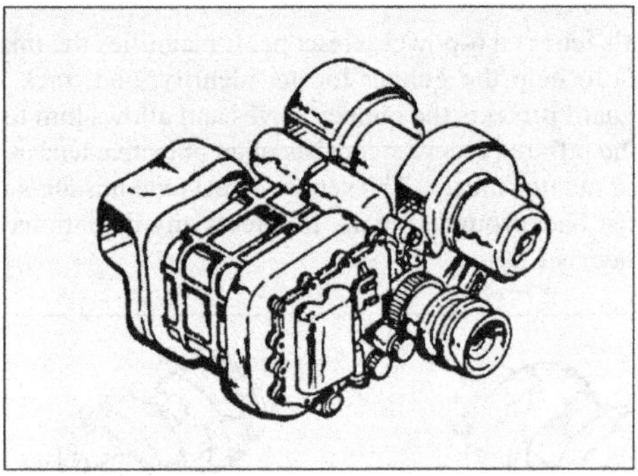

**Figure 1-5. Nightsight.**

(1) The passive, 4-power AN/TAS-5 nightsight has a 3.4-degree by 6.8-degree FOV. The nightsight is larger and heavier than the daysight. The nightsight converts heat emissions (infrared energy) from a target area into electrical signals, then to visible light (Figure 1-6, page 1-4). This enables the nightsight to display a real-time scene. Objects warmer than the ambient (surrounding) temperature appear in shades of red; cooler objects appear black. Like the daysight, the nightsight monitors the missile and sends commands to keep it on the gunner's line of sight.

(2) The firing mechanism, control system, and guidance system are the same as those on the daysight.

(3) The externally mounted 4.8-volt DC battery and coolant cartridge enable the nightsight to function without an external power source. If the appropriate accessory equipment is available, the nightsight can use military vehicle power.

(4) The nightsight has more controls and operates differently than the daysight.

**Figure 1-6. Nightsight's infrared (heat) capability.**

(a) The ACTUATOR switch (Figure 1-7) turns the system on and off. This four-position switch allows the gunner to evaluate the operating condition of the nightsight. It is normally set to the OFF-LOCK position. Rotating the switch to the AIR-BATT-CHECK position allows the gunner to check the condition of the coolant cartridge and battery. If the indicator light is off and the coolant cartridge and battery conditioner are good, the gunner rotates the switch to the ON position to operate normally.

**CAUTION**

Use the fourth position, RELEASE, to release an expended coolant cartridge.

(b) The gunner uses the focusing ring to adjust the focus. The brightness (*BRT*) and contrast (*CTRS*) controls work the same as those on a television set (Figure 1-8, page 1-5).

c. **Ammunition**. The Dragon's ammunition is an expendable component consisting of both the missile and the launcher itself (Figure 1-9, page 1-6). The missile is installed in the launcher at the factory and is shipped ready to fire. The launcher serves as a storage and carrying case for the missile before it is launched. The Dragon has two types of missiles: the M223 practice round, which has an inert warhead, and the M222 HEAT round. The launcher consists of—

- A smooth-bored fiberglass tube.
- A breech and gas pressure generator assembly.
- A wiring harness.
- A sling.
- Forward and rear shock absorbers.
- A sight battery, which provides power to operate the sight and fire the missile.
- A sight support bracket, which houses the Dragon's electrical connectors.
- A bipod, which supports the weapon system during dismounted firing.

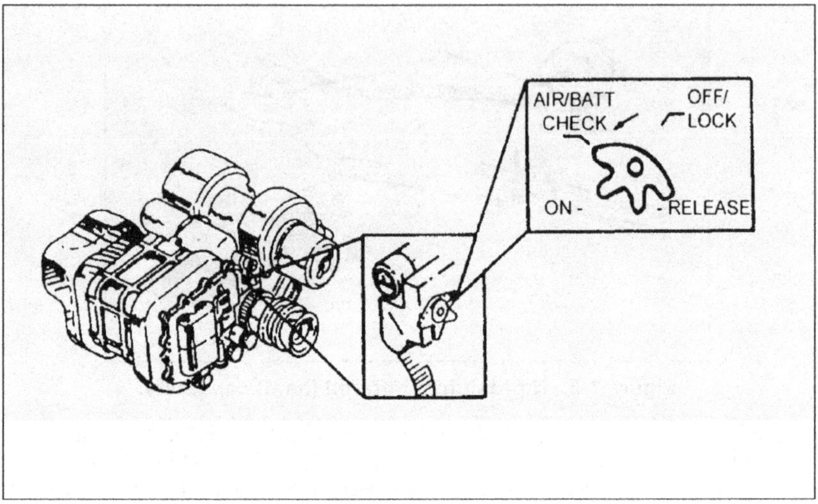


**Figure 1-7. Nightsight actuator switch.**

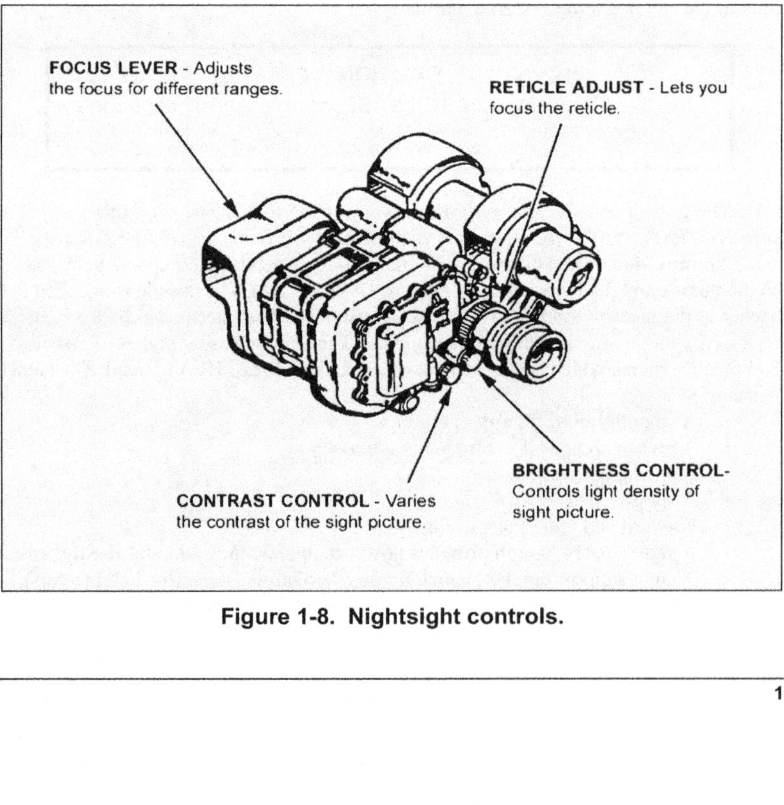


**Figure 1-8. Nightsight controls.**

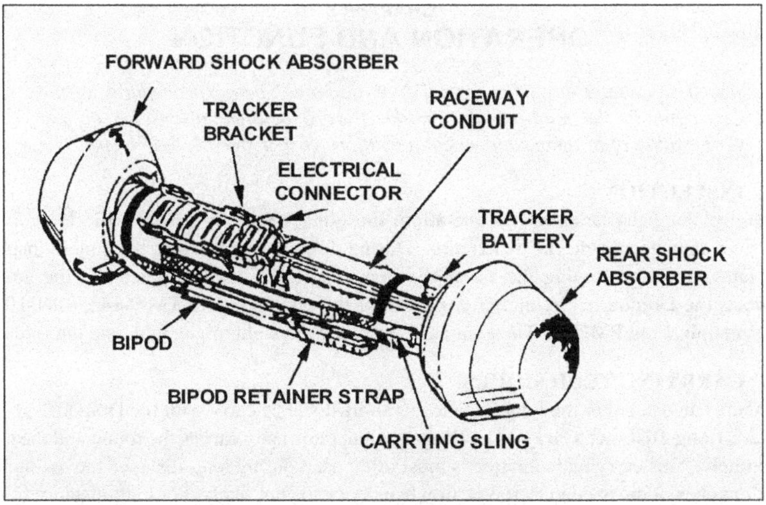


**Figure 1-9. External components of the M222 HEAT (tactical) round.**

## 1-2. TECHNICAL DATA

Table 1-1 shows technical data for the Dragon.

| | | | | |
|---|---|---|---|---|
| **AMMUNITION** | M222 HEAT round | Weight | 14.60 kg | (25.29 lb) |
| | | Missile Length | 744 mm | (29.39 in) |
| | | Launcher Length | 1,154 mm | (44.10 in) |
| | M223 inert practice round | Weight | 4.70 kg | (10.40 lb) |
| | | Missile Length | 846 mm | (33.32 in) |
| | | Launcher Length | 1,154 mm | (44.10 in) |
| **OPERATIONAL CHARACTERISTICS** | | Maximum range | 1,000 m | |
| | | Minimum range | 65 m | |
| **DAYSIGHT--**Infrared Tracker, Guided Missile, SU-36/P | | Weight | 3.10 kg | (6.75 lb) |
| | | Length | 196 mm | (7.72 in) |
| **NIGHTSIGHT--**Night Vision Sight, Tracker, Infrared, AN/TAS-5 | | Weight | 9.82 kg | (21.65 lb) |
| | | Length | 368 mm | (14.50 in) |

**Table 1-1. Technical data for the M47 Dragon.**

CHAPTER 2
# OPERATION AND FUNCTION

*This chapter discusses normal operation and function procedures for the Dragon in the man-portable mode. (For a detailed discussion on the operation of Dragon support equipment, refer to TM 9-1425-484-10.)*

## 2-1. INSPECTION

Before placing the Dragon into operation, the gunner visually inspects the Dragon for physical damage and checks its function. During offensive operations, the gunner inspects the Dragon before leaving the assembly area. During defensive operations, the gunner inspects the Dragon before and after preparing the position. (See TM 9-1425-484-10 for information about PMCS.) The gunner inspects the daysight, nightsight, and the round.

## 2-2. CARRYING TECHNIQUES

Gunners can use either the long-distance or short-distance carry with the Dragon.

a. **Long-Distance Carry**. The Dragon gunner normally carries the round and the sight separately. This carrying technique is most often used during long-distance moves such as foot marches or cross-country travel, or when contact is not likely. For long-distance travel, the gunner uses carrying technique 1, 2, or 3, as shown in Figure 2-1 (page 2-2).

b. **Short-Distance Carry**. The gunner can carry the Dragon short distances with the sight mated to the round. When using this type of carry, the gunner must keep the forward protective shock absorber and lens covers in place until the weapon is likely to be used. The protective lens covers prevent possible damage to the sight lens and keep foreign objects out of the launcher. Carry technique 4 and 5, also shown in Figure 2-1 (page 2-2), can be used with the round and the sight mated. The soldier should only use one of these carry techniques when firing is likely. Otherwise, he should avoid carrying the weapon system in a ready-to-fire configuration.

## 2-3. PREPARATION FOR FIRING

Improper or careless handling of the round can damage its components and cause the missile to malfunction when launched. If there is any reason to believe the missile has been damaged, the round must be returned to the responsible ammunition personnel for inspection and disposition. Before using the nightsight, the gunner must conduct a preoperational check IAW TM 9-1425-484-10.

a. **Releasing the Bipod**. Unsnap the bipod-retaining strap (A, Figure 2-2, page 2-3). Push the bipod forward until you feel resistance. Snap off the forward shock absorber. Ensure that the desiccant bag, attached to the forward shock absorber, is not lodged in the launch tube (B, Figure 2-2, page 2-3). While still holding the round, depress the bipod friction lock. Push the bipod legs downward to number 4 or 5, then release the friction lock (C, Figure 2-2, page 2-3).

b. **Selecting a Firing Position**. Select the firing position that best meets the situation. For training, use the sitting position (discussed later in this chapter). In a combat situation, you can use any one of the three positions: sitting, kneeling, or standing supported.

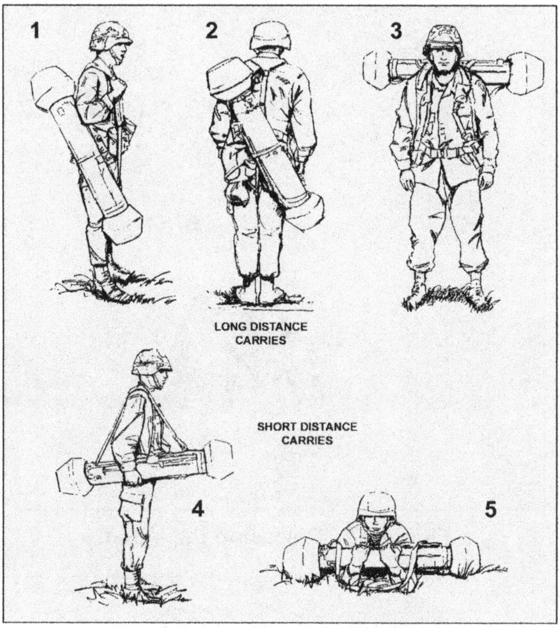


**Figure 2-1. Dragon carry techniques.**

**CAUTION**

1. Do not use the shock absorbers as handles to lift the sight. The shock absorbers might tear off.
2. Do not touch the lenses, as this can easily damage them.

c. **Removing the Sight from the Carrying Bag or Rucksack**. Open the carrying bag or rucksack. Grasp the sight by the trigger mechanism or telescoping barrel. Remove the sight and remove the cover from the electrical connector. Secure the cover to the hook-pile tape on the forward shock absorber.

d. **Removing the Lens Cover**. Secure the lens cover to the top of the forward shock absorber (Figure 2-3). Visually inspect the lenses for damage or obstruction. When using the nightsight, make sure the ACTUATOR switch is in the ON position.

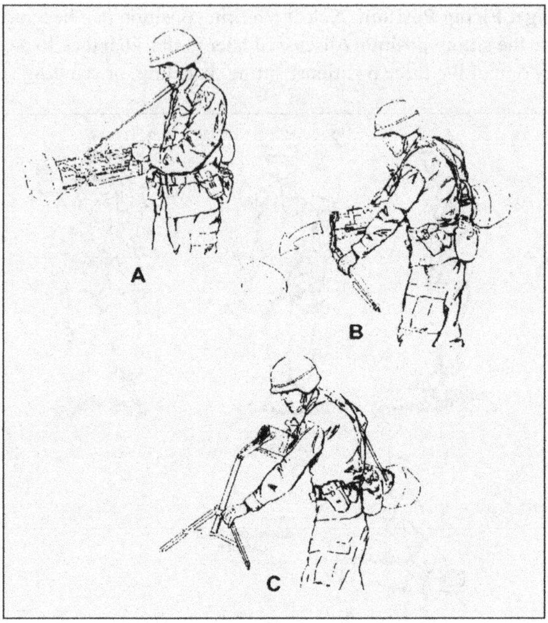


Figure 2-2. Releasing the bipod.

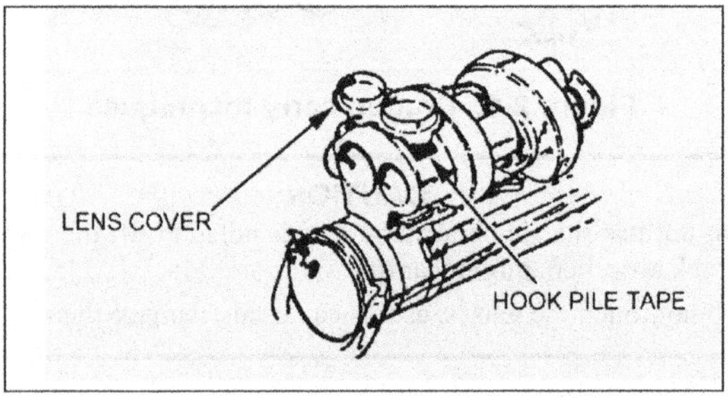


Figure 2-3. Cover secured to forward shock absorber.

e. **Mating the Sight to the Round**. Place the sight guide pins in the slots of the sight bracket guide rails (Figure 2-4, page 2-4). This example uses the daysight, but you will use the same procedure for the nightsight. Use both hands and slide the sight firmly to the rear until the spring clip locks the guide pin in place (Figure 2-5, page 2-4). If the sight does not mate, notify your squad or section leader. Assume the firing position and put the round on your shoulder.

**WARNING**

**Avoid pressing the safety and the trigger while mating the sight to the round.**

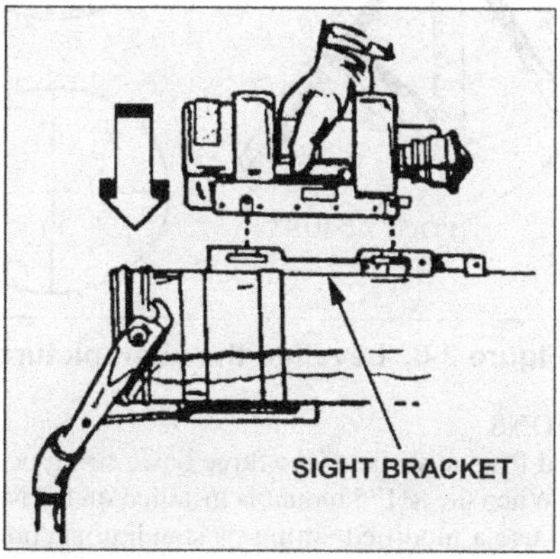


**Figure 2-4. Aligning the daysight guide pins.**

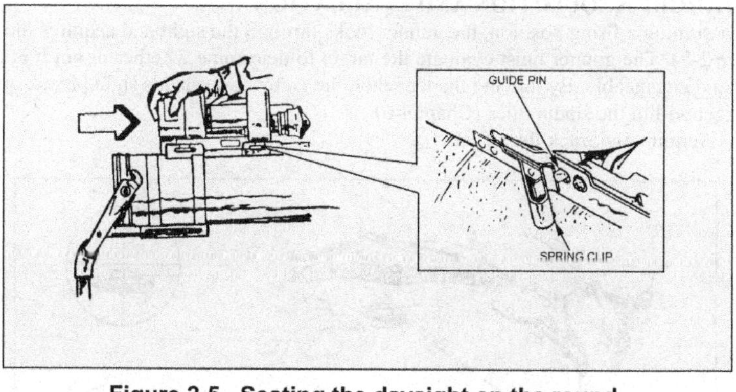


**Figure 2-5. Seating the daysight on the round.**

f. **Adjusting for Height and a Level Sight Picture**. Look through the sight to see if the sight picture is level. An unlevel sight picture (more than 6 degrees off level) will cause the sight to send bad commands to the missile. The missile will fly erratically and the missile will hit the ground. To obtain a level sight picture, adjust the bipod friction lock and the foot adjustment lever (Figure 2-6, page 2-5).

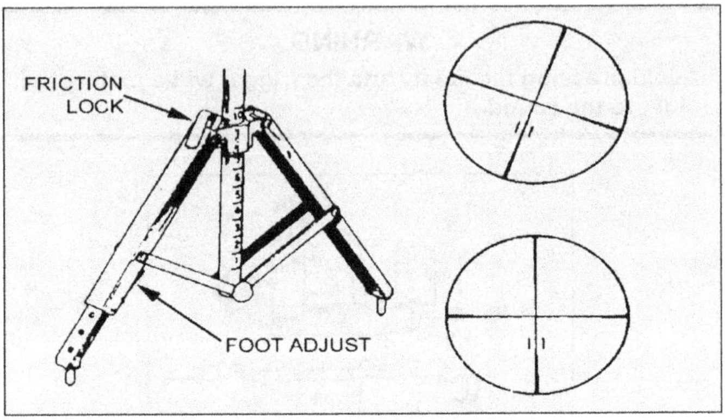


Figure 2-6. Leveling the sight picture.

## 2-4. FIRING POSITIONS

The Dragon can be fired from any one of the three basic firing positions: sitting, standing supported, or kneeling. When the M175 mount is installed on the M3 or M122 machine gun tripods, the gunner can use a modified sitting or standing supported position. Chapter 5 discusses firing positions in detail.

## 2-5. TARGET ACQUISITION AND EVALUATION

After assuming a firing position, the gunner looks through the sight and acquires the target (Figure 2-7). The gunner must evaluate the target to determine whether or not it is within range and engageable. By moving the launcher, the gunner adjusts his sight picture, placing the target within the stadia lines (Chapter 6).

a. Acquire and track the target.

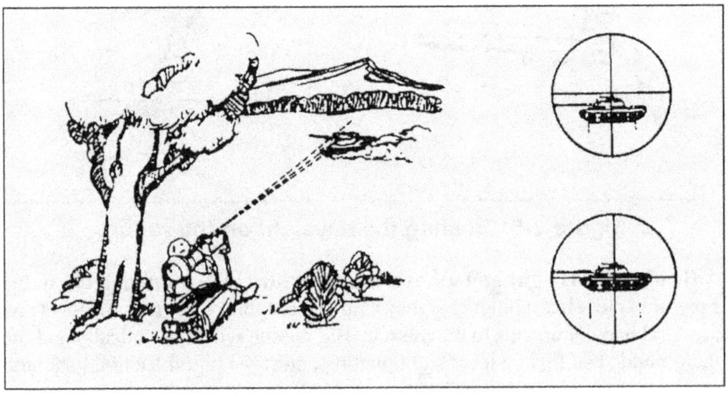

Figure 2-7. Target acquisition.

b. To fire the weapon, push the safety plunger in with the thumb of the right hand, depress and hold the trigger, and maintain the sight picture until impact (Figure 2-8). When engaging a moving target, move the upper part of your body left or right to adjust for azimuth; move up and down to adjust for elevation. Jerking, bumping, or moving the sight while the missile is in flight may cause you to miss the target. To establish a smooth tracking rate, track the target for about two seconds before squeezing the trigger.

c. Ignore the missile when it appears in the sight picture. Do not try to fly the missile. Keep the cross hairs on the target—let the sight guide the missile.

**Figure 2-8. Correct sight picture.**

## 2-6. MISSILE FLIGHT AND CONTROL

As the missile leaves the launch tube, the infrared flares begin to operate, and the command-link wire is dispensed from the missile. The missile fins fold out and lock into place. The safety and arming device sequence is completed after the missile has traveled about 65 meters (Figure 2-9).

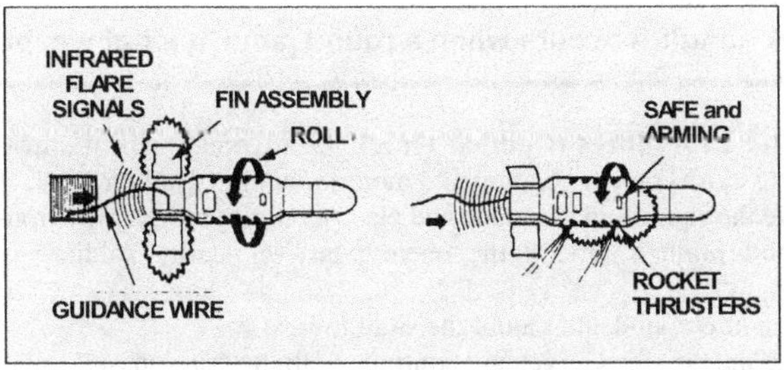


**Figure 2-9. Dragon functioning concept.**

a. The sight's electronic controls compare the missile's position with the gunner's LOS. It sends electronic position information to the missile through the command-link wire.

b. Based on information received from the sight, the missile generates commands to correct its position. The missile may store the commands until it rolls to the proper position for its rocket motors to fire, or it may fire the rocket motors immediately to correct the missile's position relative to the gunner's LOS.

c. The missile's rocket motors fire to accelerate it or to correct its position. The gunner keeps the cross hairs on the selected aiming point until impact.

d. To fire a second round, the gunner releases the spring clip at the rear left side of the sight support assembly, slides the sight forward, and lifts it clear of the support. The gunner expends or destroys and discards the round IAW unit SOP. The gunner either prepares another round or he secures the sight and any extra rounds and prepares to displace on order.

## 2-7. FAILURE TO FIRE

A failure to fire occurs whenever the missile does not leave the launcher after the gunner depresses the trigger lever bar. When the missile fails to fire, the gunner immediately resqueezes the trigger lever bar and continues to track the target for 15 seconds. If after 15 seconds the missile has not fired, the gunner announces "Misfire," releases the trigger lever bar, and carefully feels near the sight battery. If the sight battery is hot, the gunner performs hangfire procedures; if it is cold, the gunner performs misfire procedures. Ideally, a two-man team operates the Dragon, and the assistant gunner secures the new round. In the normal tactical situation, a moving armored vehicle closes on the position and moves to within 1,000 meters. To preserve forces, leaders should have an emergency action SOP in place that covers what to do if a Dragon fails to fire.

**WARNING**

**A hangfire is a delayed firing. It is difficult to distinguish immediately from a misfire. The difference lies in when they occur and what causes them.**

- **A misfire occurs when a gunner tries to fire a round from a cold weapon and the round jams for mechanical reasons.**
- **A hangfire occurs when a round jams in a hot weapon.**

a. **Hangfire Procedures (Combat Only).** In a typical tactical situation, a moving armored vehicle closes on the position and moves to within 1,000 meters.

(1) Remove the sight from the round and place it on the ground away from the firing site. Keep the round pointed toward the enemy; advise nearby soldiers of the round's failure to fire.

(2) Obtain a new round and mount the sight to it.

(3) Acquire the previous target and continue with the mission.

b. **Hangfire Procedures (Training Only).**

(1) The gunner notifies the range officer or NCOIC about the hangfire.

(2) The gunner removes the round from his shoulder and places the round flat on the ground. The gunner ensures the bipod legs are pointing to the side, with the round pointing downrange.

(3) The gunner moves away from the position.

(4) The range officer or NCOIC clears all personnel within 50 meters of the round and notifies the EOD.

c. **Misfire Procedures.**

(1) Loosen the sight from the round, then remate it to the round.

(2) Track the target and try to fire again.

(3) If the round again fails to fire, feel carefully near the sight battery once more.

(4) If the battery still feels cold, remove the sight from the round. Place the round on the ground away from the firing site. Keep the round pointed toward the enemy; advise soldiers near you of the misfire.

(5) Mount the sight on another round and continue with the mission.

(6) If the new round also fails to fire, carefully feel near the sight battery. If that battery is also cold, the sight is probably defective.

(7) If you can find another sight, replace the existing sight and continue with the mission, using previously unfired rounds.

## 2-8. RESTORATION TO CARRYING CONFIGURATION

The gunner uses a long-distance or short-distance carry to move a Dragon round that he has prepared for firing but that he no longer needs. To use the long-distance carry, the gunner performs Steps 1 and 2. To use the short-distance carry, the gunner performs only Step 2.

a. **Step 1.** Remove sight from round.

(1) Replace sight lens cover.

(2) Remove sight from round; and replace connector cover on sight and round.

(3) Return sight to carrying bag or rucksack.

b. **Step 2.** Restore round to carrying configuration.

(1) Replace connector cover.

(2) Replace forward shock absorber.

(a) Set round in an upright position with rear shock absorber resting on the ground.

(b) While lightly pushing down on bipod, press bipod brace toward round to lift it from locking slot.

(c) Once bipod brace is unlocked, push down on bipod until upper part of bipod that attaches to round is parallel with muzzle of launcher.

(d) Align cutout portion of forward shock absorber with bipod at launcher muzzle.

(e) Push down on shock absorber while lowering bipod against round. Ensure upper part of bipod engages shock absorber.

(f) Retract bipod legs and secure them to round with retainer strap.

(3) Lift up on forward shock absorber to ensure it is secure. Round is now in safe carrying configuration.

## 2-9. ADVERSE WEATHER CONDITIONS

If the gunner can see the target, the Dragon can operate in all weather conditions. The Dragon and round stay the same temperature. For example, if the gunner removes a Dragon from a vehicle whose inside temperature is 45 degrees Fahrenheit, both the Dragon and the round will measure 45 degrees Fahrenheit, even if the outside temperature measures 32 degrees Fahrenheit.

a. **Cold Weather**. Before deployment, check all optical surfaces and all mechanical devices for proper operation. Ensure there is neither ice nor snow on the sight components. Never operate the Dragon in temperatures below minus 25 degrees Fahrenheit.

b. **Hot Weather**. To avoid subjecting the round and sight to prolonged, direct sunlight, provide as much shade or cover as possible. Never operate the Dragon in temperatures above 145 degrees Fahrenheit.

CHAPTER 3
# TRAINING EQUIPMENT

*Training equipment provides units with a low-cost means to develop and maintain gunnery skills. To train effectively, trainers must know the weapon and the training equipment. This chapter describes Dragon training equipment and provides a program to train instructors to use it. (Appendix A provides train-the-trainer tasks for the DGT and DFTT.)*

## 3-1. PRECISION GUNNERY TRAINING SYSTEM

The Dragon PGTS includes one indoor training system and one outdoor training system. The Dragon gunnery trainer (DGT) is the indoor training system. It provides a means for precision gunnery training, practice, and qualification with the Dragon. The Dragon field tactical trainer (DFTT), the outdoor training system, provides a means for familiarization with the Dragon. (Appendix B and Appendix C discuss the DGT and DFTT, respectively.)

a. The realistic training offered by the Dragon PGTS allows units to achieve a high level of proficiency without firing live Dragons. Individual units are responsible for conducting sustainment training on the Dragon PGTS.

b. Soldiers can obtain instructional videotapes on the Dragon PGTS from training support centers (TSCs). Basic instructional fire (Firing Tables 1 through 4) is conducted on the DGT. These tables introduce new gunners to the basic gunnery skills required to engage targets with a Dragon. Firing Tables 5 and 6 are conducted in a simulated outdoor tactical situation using the DFTT. Firing Tables 7 and 8 are conducted on the indoor DGT. Firing Table 7 allows gunners to practice for qualification. If they score 16 out of 20 on Firing Table 8 (Table 3-1), gunners qualify on the Dragon. (Appendix D describes firing tables and provides examples of completed firing tables.)

| FIRING TABLE NUMBER | PURPOSE OF FIRING TABLE | TRAINING DEVICE | PAGE IN MANUAL |
|---|---|---|---|
| 1 | Familiarization | DGT | D-1 |
| 2 | | | |
| 3 | | | |
| 4 | | | |
| 5 | Familiarization | DFTT | D-2 |
| 6 | | | |
| 7 | Practice for qualification | DGT | D-3 |
| 8 | Qualification (Hit 16 out of 20) | | |

**Table 3-1. Firing tables for Dragon training devices.**

c. **Train-the-Trainer Program.** A successful Dragon training program requires well-trained NCOs who know and have confidence using both trainers. Each company should have at least one certified trainer. (Appendix A provides train-the-trainer tasks for both the DFTT and the DGT.)

## 3-2. DRAGON GUNNERY TRAINER

The DGT man-portable trainer simulates the Dragon guided missile system's sight(s), controls, switches, and indicators. (See Appendix B for a complete description of the DGT and its components.) Training programs selected by the training NCO display simulated battlefield scenes through the sight. These scenes include both enemy and friendly vehicle targets. The indoor, 110- or 220-volt AC Dragon gunnery trainer enables units to train personnel on the Dragon without a range or training area (Figure 3-1).

a. **Role**. Soldiers train on the DGT to—

- Assume the correct Dragon firing position.
- Identify targets.
- Determine engageability of targets.
- Engage targets (including tracking and firing).

b. **Missions**. Matched sets of videodiscs and floppy (data) disks provide a library of missions. Each disk set contains 30 missions of various lengths plus learning objectives. Each mission applies to both daysight and nightsight viewing.

c. **Target Options**. The DGT allows the trainer to adjust the difficulty of the scenario to the shooter's skill. To do this, trainers can—

- Change the target kill zone.
- Change the obscuration.
- Vary the amount of weight the system loses after a round is fired.
- Select night fire.
- Select multiple or moving targets.

**Figure 3-1. Dragon gunnery trainer.**

d. **Feedback**. The DGT enables the trainer to provide visual feedback to the firer in either of two ways. He can replay the mission on video, or he can print out two views (from the top and side) of the actual mission. Both show horizontal and vertical tracking errors and the firer's score. Figure 3-2 shows the instructor's console.

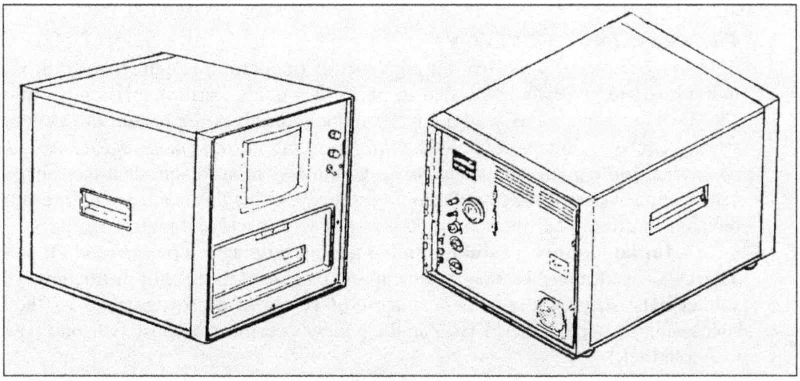

**Figure 3-2. Instructor's console, Dragon gunnery trainer.**

### 3-3. DRAGON FIELD TACTICAL TRAINER

The DFTT outdoor trainer realistically simulates the appearance, feel, launch, flight, and effects of the Dragon guided missile system. (Appendix C describes the DFTT and its components.) Training with the DFTT can occur on designated ranges, general outdoor areas, or representative tactical environments. The DFTT uses the Dragon daysight or nightsight and trainers can use it instead of the inert practice round. The DFTT is bipod-mounted, just like an actual Dragon.

a. **Role**. Trainers use the DFTT to familiarize gunners with firing the Dragon in an outdoor environment. Specifically, they use it to train soldiers to—

- Assume the correct Dragon firing position.
- Determine engageability of targets (whether targets can be engaged).
- Engage targets (including tracking and firing).

b. **Targets**. Each target has a retroreflector. This acts as a mirror and returns part of the laser beam sent by the trainer round. The DFTT transmits Dragon MILES code, so it can also kill targets equipped with a MILES (multiple integrated laser engagement system) harness, as long as the target also has a retroreflector.

CHAPTER 4
# UNIT TRAINING

*This chapter guides the unit leader in developing a successful unit training program.*

## 4-1. TRAINING STRATEGY

A training strategy is the overall concept for integrating resources to train the skills—individual and collective—needed to perform a unit's wartime mission. Commanders at TRADOC institutions as well as in the units themselves implement the Dragon training strategy, whose four parts include *initial training*, *sustainment and advanced training*, *collective training*, and *leader training*. Commanders make sure that they include trainer certification as part of the overall training strategy. Using this manual, a unit commander can develop an effective Dragon training program that meets unit requirements.

a. **Initial Gunner Training**. Initial gunner training is a prescribed POI conducted in TRADOC institutions. Initial gunner training includes 15 blocks of instruction. To pass, the soldier must successfully achieve a score of 16 out of 20 engagements on the DGT and successfully complete the 11 tasks in the gunner's performance test. (Chapter 6 provides an example POI.)

b. **Sustainment and Advanced Training**. Sustainment and advanced training are conducted in the unit.

(1) ***Sustainment Training***. Sustainment training, conducted monthly, quarterly, and annually, ensures gunners retain the skills they learned in initial training. To retain their skills, gunners practice precision tracking on the DGT at least monthly. To practice, they fire selected engagements from the monthly sustainment table using the DGT and DFTT. They must also complete at least 50 percent of the gunner's skill test. (Chapter 7 discusses sustainment-training programs in detail.)

(2) ***Advanced Training***. The unit conducts advanced training. It consists of tracking exercises using the DGT, DFTT, or the Dragon MILES. The tracking engagements grow more difficult. Existing conditions, equipment, or tactical play determine difficulty. Examples of advanced training include—

- Night tracking exercises using artificial illumination or the AN/TAS-5 nightsight.
- MOPP tracking exercises.
- Other situational gunnery exercises.

c. **Live-Fire Training**. The number of live missiles the gunner may fire depends on the type of unit and on budget constraints. The unit may conduct annual live fire in an instructional setting or may integrate it into other live-fire exercises. Where possible, live missile firings should closely follow a scheduled qualification. Only currently qualified gunners should be allowed to fire live missiles.

d. **Collective Training**. The unit conducts collective training to fully integrate the Dragon into the unit's overall combat power. Collective training has two parts: force-on-force training and live fire training.

(1) The unit conducts *force-on-force training* with MILES during a squad and platoon FTX or STX. During semiannual external evaluations, the unit evaluates the platoons on Dragon employment.

(2) The unit conducts *live-fire training* using the laser target interface device (LTID) or actual missiles (live or inert) along with platoon LFXs. Dragon gunners or teams should participate in squad or platoon collective live-fire exercises at least twice a year.

e. **Leader Training**. Leader training occurs in the unit through NCO and officer development classes and personal initiative. Leaders selected as Dragon instructors must have already *demonstrated* proficiency in using the Dragon and should periodically recertify to maintain leader proficiency.

## 4-2. COMMANDER'S RESPONSIBILITY

The commander plans, executes, and supervises training, selecting instructors and gunners based on the criteria in this chapter. He makes sure that instructors and gunners receive training. The commander does the following:

a. Allows adequate time for effective training.

b. Ensures all training meets training standards.

c. Combines Dragon training with other unit tactical training.

d. Ensures the maintenance facility properly maintains all consolidated Dragon training equipment.

e. Conducts sustainment training.

f. Ensures subordinate leaders conduct operational readiness checks.

g. Periodically inspects all tactical and training equipment.

h. Keeps a formal record of—

- Inspections.
- Training results.
- Gunner qualifications.
- Gunner turnover.

i. Evaluates the Dragon training and corrects deficiencies in future training.

## 4-3. TRAINING PROGRAM GOALS

Training program goals provide leaders with the objectives necessary for success. For soldiers to achieve proficiency in Dragon gunnery, leaders must—

- Train them to successfully engage armored targets.
- Train them to maintain and operate the Dragon and related training devices.
- Provide clear sustainment training standards.
- Make sure gunners know how to meet the standards.
- Train gunners to those standards.
- Conduct sustainment training.
- Evaluate each gunner's performance.

## 4-4. INSTRUCTOR SELECTION

Dragon instructors have to meet the same standards that gunners do. The commander picks soldiers who want to and can instruct and who have a high level of tactical competence at squad and platoon levels. Experience as a Dragon gunner will serve an instructor well, but does not guarantee a soldier's success as an instructor. Finally, the overall training strategy must include trainer certification.

## 4-5. GUNNER SELECTION

First, the leader identifies soldiers who want to be Dragon gunners. Then, he makes sure each soldier closely approaches the physical standards for Dragon gunners shown in Table 4-1.

1. To track moving targets, the soldier must be able to flex his upper trunk to the left and right.
2. To hold the weapon steady during tracking and to carry it as required, a soldier must have considerable upper body strength and physical stamina
3. To use the Dragon, a soldier must stand between 5 feet tall and 6 feet 2 inches tall—the bipod only adjusts for this height range.
4. To acquire a target, fire the Dragon, and track the round, a soldier must show that he can hold his breath for at least 13 seconds.
5. To use the Dragon, a soldier must have unaided minimum vision of 20/100, correctable to 20/20. The focus (reticle adjustment) can only compensate for this amount of near-sightedness.
6. To use the Dragon, a soldier must have right-eye dominance. He must be able to close his left eye independently. That allows him to keep his left eye closed to protect it from distracting dust and debris.

**Table 4-1. Physical standards for the Dragon gunner.**

## 4-6. MATCHING THE TRAINING PROGRAM TO THE UNIT

Mandatory training programs seldom fit a unit's particular circumstances and needs. Only the unit's own commander can develop its training program. Table 4-2 shows the steps for developing a unit-training program. The key is to consolidate training at unit level. For this to succeed, the commander must—

- Assign a qualified NCO with additional duties as NCOIC of Dragon training.
- Include Dragon training on both short-range and long-range training calendars.
- Hold subordinate leaders accountable for execution of training.

## 4-7. LEADER TRAINING

Each commander must determine what specific, Dragon-related tasks he and other leaders must perform. Then he develops these tasks into training objectives. Next, instructors choose proper training methods, depending on the resources available and other training requirements. If incoming leaders do not know how to employ the Dragon, the commander makes sure they receive Dragon training—from the basics on up through the more difficult tasks. Initial Dragon training works well in the classroom, since it depends on the use of lectures, seminars, small discussion groups, and briefings. This cuts down on the use of expensive training resources such as ranges and transportation.

| |
|---|
| 1. Identify all the unit's tactical and administrative missions. |
| 2. Analyze the unit's missions. Determine what individual and unit tasks soldiers must accomplish in order to complete the mission. |
| 3. Establish individual and unit training objectives to accomplish the unit's tasks. |
| 4. Determine the level of individual and unit proficiency in the unit's tasks. |
| 5. Determine individual and unit training needed to attain the training objectives. |
| 6. Identify available training resources. |
| 7. Program and schedule training based on the training resources available and on individual and unit training needs. |
| 8. Conduct training. |
| 9. Monitor and evaluate training, and revise the training program as required. |

**Table 4-2. Steps in developing a unit training program.**

### 4-8. GUNNER TRAINING

Dragon gunners should be trained consistently and as repetitively as resources allow. All training includes gunnery and night fire.

a. **Training Methods**. Two main methods are used to train the Dragon gunner:

(1) ***Centralized Gunnery Training***. Centralized gunnery training requires the unit to train all Dragon gunners at once. Training type depends on the training resources available (equipment, facilities, personnel, and time) and the tasks to be taught.

(a) *Equipment and Facilities*. Limited distribution of training equipment suggests centralized control of training at the highest unit level possible. Also, the number of firing ranges suitable for the Dragon may dictate centralized control over these facilities.

(b) *Personnel*. A shortage of qualified Dragon instructors can hamper gunnery training.

(c) *Time*. Centralizing Dragon training can save training time since fewer instructors and classes are required. For example, if the training is centralized at brigade level or higher, battalion and company level instructors can focus on other training requirements. With centralized training, qualified Dragon instructors prepare and conduct training. They need less time to prepare than would unqualified instructors.

(d) *Tasks*. When considering centralized training, the commander also considers the task he wants to train. If the task relates to gunnery training or qualification, the number of training sets and ammunition available might limit the choice to centralized training. Dragon gunners first learn how to prepare a basic range card. Then they practice applying what they have learned using different pieces of terrain or terrain substitution (maps, sand tables, or 35-mm slides of the terrain). Gunners build these skills by working in small groups where they can ask questions, talk about the answers, and debate the advantages and disadvantages of the range cards.

(2) ***Round-Robin Training***. Battalions have a limited number of training sets, so soldiers cannot practice all at once. Rather than letting a few soldiers practice on the equipment while others watch, trainers can set up round-robin (multistation) training. Soldiers then rotate through the stations. This keeps everyone actively engaged in training the whole time. For example, soldiers might train on the equipment itself at Station Number 1, learn to prepare range cards at Station Number 2, and learn to identify enemy vehicles at Station Number 3 (Figure 4-1).

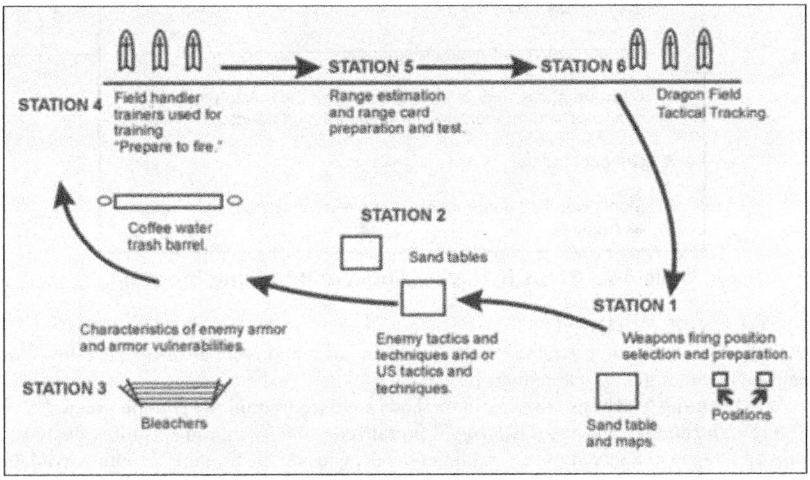


**Figure 4-1. Example of round-robin training.**

b. **Gunnery**. Gunnery qualifies gunners to fire the Dragon tactically and nontactically. To qualify, gunners must meet the following objectives:

(1) Detect vehicles at different ranges under varying field conditions, such as rolling hills and vegetation, with the vehicles moving and stationary.

(2) Determine whether a moving target, if engaged, will reach cover before impact.

(3) Prepare a firing position and range card.

(4) Know how to lessen the signature of the backblast.

(5) Know what suppressive fires the enemy can place on the firing position.

(6) Know how to use cover and concealment, deception, surprise, and movement.

(7) Know unit SOPs covering rules of engagement, including signals to lift or shift fires, priority of targets, and times to engage targets.

(8) Know where to obtain resupply of missiles.

(9) Know how to inspect the round before firing.

c. **Night Training**. Since threat doctrine stresses night operations, gunners should practice their skills at night. They can do this during FTXs, with the gunners using the FHTs with nightsights, or they can do it on a range, under controlled conditions, with artificial illumination.

## 4-9. TEAM TRAINING

Due to the frequent need for leaders to employ Dragons as part of a team, team training must become an integral part of Dragon training in the field. As a rule, Dragon gunners employed apart from their squad positions have at least one assistant gunner and ammunition bearer with them, or they may be employed in an antiarmor fire team. The other members of the team work with the gunner as a coordinated element. As such, their duties include providing security for the gunner, helping the gunner prepare his firing position, carrying ammunition (missiles), and locating targets.

a. Once gunners have qualified with the Dragon, and once leaders have trained in tactical employment of the Dragon, they can train together profitably in FTXs.

(1) During their training, leaders must decide how to employ the Dragon. They could use scenarios, simulated threat conditions, or OPFOR, for example. Based on these simulated conditions, platoon leaders practice choosing armor avenues of approach, selecting Dragon firing positions, and making related employment decisions.

(2) Leaders should have gunners prepare firing positions and range cards, and simulate engaging targets. Employing gunners with their squads helps train the other squad or fire team members to perform their duties (providing security, locating targets, and so forth). Then, leaders should evaluate and discuss the training exercises to decide if the techniques and tactics used were the best ones for the situation.

b. When qualifying on the range, Dragon gunners usually learn under near ideal conditions: They have excellent fields of fire and a target that moves conveniently back and forth on level ground directly in front of them. Except when other gunners fire, they have no distractions. So, after qualifying on the range, they are ready for more advanced training. This means using the Dragon under more realistic, and therefore more difficult, conditions, which the commander creates. The following paragraphs suggest a few ways to add realism to Dragon training:

(1) During both FTXs and practice firing exercises on the range, train the gunner to concentrate while tracking a target. Try to distract him with various combinations and amounts of smoke, haze, and harassing fires. Detonate explosives near him to simulate the noise of enemy artillery and tank fires. To simulate nearby explosions, detonate explosives under a bag of flour. Obscure the target with smoke in or near the gunner's position or target—this makes tracking more realistic.

(2) During MILES-Dragon tracking exercises, the target vehicle with the MILES receivers must travel over various terrain. This moves the target across the front or towards

the gunner's position over various terrain such as curving roads, or wooded or rolling terrain.

## 4-10. COLLECTIVE TRAINING

In addition to monthly or quarterly training and qualification (or both), Dragon gunners must demonstrate their proficiency during squad and platoon exercises. Each Dragon gunner takes part in live-fire exercises as part of a squad or larger unit. He does the same in a platoon external evaluation based on standards in ARTEP 7-8-MTP (RDL version). In live-fire exercises, the unit uses actual Dragon rounds or LTIDs with other small-arms targets. Platoon external evaluations should include medium antiarmor weapons tasks and should be conducted as part of a company FTX or STX.

## 4-11. EVALUATION

One of an instructor's major responsibilities is to evaluate training. The training to be evaluated may be training that the instructor conducted or training that was conducted by another instructor. The evaluation is concerned with the *effectiveness* and *efficiency* of training.

a. *Training effectiveness* refers to whether the soldiers, teams, or units met the standards established in the commander's training objective(s).

b. *Training efficiency* refers to how well instructors used the available training resources. Leaders should write all evaluations in the AAR format. This helps make sure that gunners remember what they learned from their mistakes. It also helps trainers improve future training.

CHAPTER 5
# INSTRUCTOR'S GUIDE TO EFFECTIVE TRAINING

*This chapter provides the instructor with the information needed to train and sustain Dragon gunners and to train soldiers as team members.*

## 5-1. TRAINING PROGRAM SCHEDULE

The number of firing sites that can be staffed with instructors and equipped with DGTs and DFTTs dictates the number of gunners for each class. This allows for the effective use of time.

a. Instructors must allow sufficient time to orient students on the equipment. Negative results occur when the training schedule is shortened. The more information each soldier receives about how the weapon and training equipment operate and on the purpose of each, the better the soldier will perform as a gunner.

b. Initial marksmanship training, in the institution or unit, teaches essential skills and develops fixed and correct procedures in marksmanship before range practice begins. Thoroughly instructing and carefully supervising practice in the initial phase saves time and ammunition during range firing. It also allows students to develop techniques and procedures necessary for well-trained Dragon gunners and teams.

## 5-2. INDIVIDUAL INSTRUCTION

To prepare for live firing the Dragon, each gunner must learn various skills and habits. Although the Dragon training equipment closely simulates firing the Dragon weapon, gunners require personalized individual instruction to develop proper gunnery techniques and procedures. Instructors should key on the following points:

- Coach and stress the gunnery techniques shown in Table 5-1.
- Use and emphasize the sitting position and standing supported positions.
- Have gunners practice tracking in both directions—left to right and right to left.
- Have gunners clean the training equipment at the end of the day.

| |
|---|
| • Tight eye contact with the eyepiece. |
| • Proper position. |
| • Steady hold at the launch. |
| • Steady tracking at all times, especially through smoke. |
| • Slow and steady aim point correction to the target. |

**Table 5-1. Gunnery techniques.**

## 5-3. TRAINING SEQUENCE

The Dragon gunner should receive training in the following sequence to control the missile launch and flight:

- Positions.
- Sighting, aiming, and firing.

- Breathing.
- Tracking exercises.
- Qualification and verification.

## 5-4. FIRING POSITIONS

The gunner must acquire and maintain a stable body position relative to the weapon and be able to move smoothly when tracking a moving target. The round must be solidly anchored on the muscle of the gunner's shoulder. His arms and hands must be properly placed to squeeze the trigger and to maintain the stability of the round. The position of the eye against the eyepiece is critical. Keeping the eye firmly pressed into (against) the eyepiece reduces launch-induced movement and prevents obscuration. The three basic firing positions for the Dragon are the sitting, standing supported, and kneeling. The gunner uses a modified sitting position to fire the Dragon from the M175 mount. The M175 mount fits the M3 or M122 machine gun tripod. (For more information about use of the tripod, see Appendix E.)

a. **Sitting Position**. Demonstrate the sitting position, then have the soldiers assume the position also (Figure 5-1). Instruct them as follows:

(1) The sitting position is the most stable. Sit with your legs extended as far as possible. Place the notch of your boot heels on the bipod and push outward.

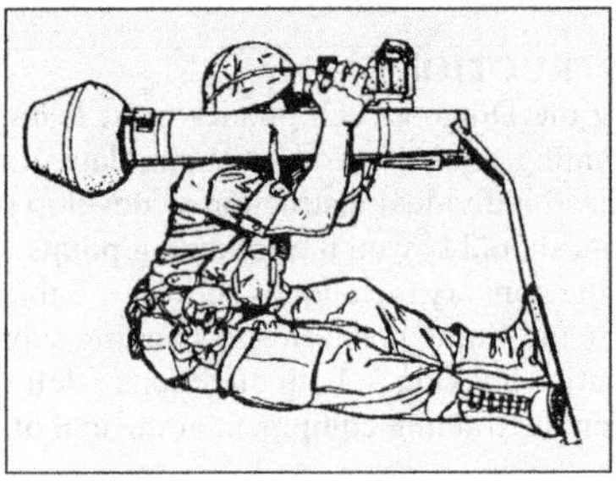

**Figure 5-1. Sitting position.**

(2) Lean forward from your waist as far as possible. Pick up the round and place it on the muscle portion of your shoulder, keeping it tight against the curve of your neck (Figure 5-2).

**Figure 5-2. Position of round on shoulder.**

(3) Grasp the barrel of the sight with your left hand, curling your thumb under the tube. Grasp the firing mechanism with your right hand, with your thumb on the safety, three fingers on the firing lever, and your little finger on the front of the firing mechanism. Place the heel of your hand on the base of the firing mechanism to provide a firm grip and reduce slippage. When firing, *hold* the trigger in the depressed position; if you release your hand, you will experience an involuntary muscle reaction that will affect the sight and in turn the path of the round.

(4) Lift your head to align your right eye with the telescopic sight. Press your head forward and press your eye firmly against the eye guard. This forces your eye to stay open. Close your left eye, and keep it closed. If necessary, focus the sight.

(5) Pull down and back with your hands while pushing out with your feet. Try to touch your elbows together and to your chest at the same time (Table 5-2).

(6) Keep your back as straight as possible while leaning forward for better breath control. This limits discomfort and increases your ability to move your upper body.

(7) Maintain arm, back, and leg muscle tension. Use enough force that you do not experience an involuntary muscle reaction when the weight of the missile is removed from your shoulder.

| |
|---|
| 1. Stress the need for keeping the pull-down force on the sight and the eye tight in the eye guard. |
| 2. Body position and breath control are the two key elements to effectively engage targets. |
| 3. Ensure the gunner keeps his body and limbs clear of the backblast area. He must keep the round at least 6 inches off the ground in order for the missile fins to clear the ground. |

**Table 5-2. Training notes.**

b. **Standing Supported Position.** Demonstrate the standing supported position, then have the soldiers assume it also (Figure 5-3, page 5-4). Instruct them as follows:

(1) While standing in an individual fighting position or behind a support, place the bipod legs to your front. Place it at such a distance that you must reach for the round.

(2) Spread your legs a comfortable distance apart, keeping them straight. Place the round on your shoulder muscle.

(3) Lean forward against the wall of the fighting position to support your body from the waist down, so that you are in a stable firing position.

(4) Grip the sight as you did in the sitting position. Pull backward and down, while straightening your upper body slightly; this removes any slack in the bipod.

(5) Place your upper body, arms, hands, head, and eyes in the same position as you did when you assumed the sitting position.

c. **Kneeling Position.** Kneel, and spread your knees a comfortable distance apart. Position the bipod so that you have to lean forward to position your eye in the eye guard (Figure 5-4). Grasp the sight as previously described. Assume the same upper body position you did for the other firing positions. Place the round on your shoulder muscle, keeping it tight against your neck. As you lower your buttocks to your heels, take the slack out of the bipod. Try to sit on your heels so you will have a stable firing platform.

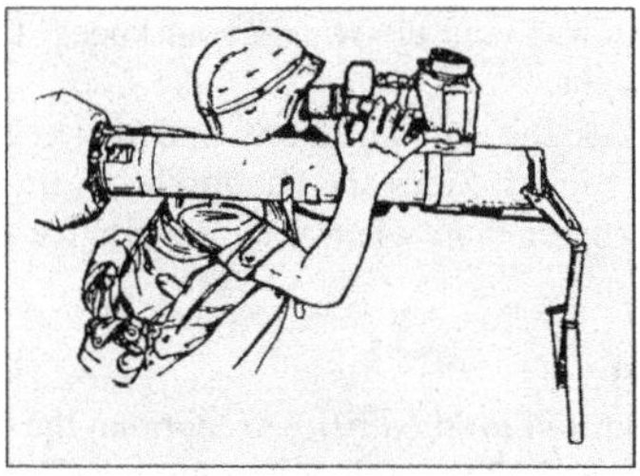

Figure 5-3. Standing supported position (fighting position).

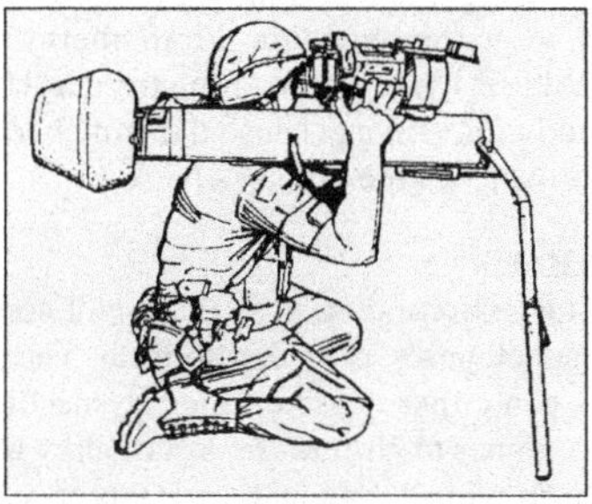

Figure 5-4. Kneeling position.

## 5-5. M113 INSTRUCTIONAL TECHNIQUES

Certain instructional techniques apply only to M113-equipped units. (See Appendix E for more information.)

## 5-6. SIGHTING, AIMING, AND FIRING

Once the gunner has mastered the firing positions, the instructor can teach him how to sight, aim, and fire properly.

a. **Sighting**. Place the eye in the eyecup and pull the weapon tight enough against your eye that you cannot blink. Wrap your small finger around the front of the firing mechanism to add pulling force and to help keep the weapon tight against your eye. Visually select a target, and then acquire it through the daysight or nightsight by adjusting the upper portion of your body.

(1) Keep the weapon tight to keep the sight picture.

(2) Do not move your eye in the eyecup. If you move your eye, you will see the side of the telescope prism, which will blur your sight picture.

b. **Aiming**. Confirm that the target is within range by using the stadia lines. Place the cross hairs of the sight on the center of the target's visible mass. Regardless of the target's range or speed of movement, keep the cross hairs on the firing point until after you have fired and observed the impact of the round. To maintain the proper sight picture on a moving target, you have to move the upper portion of your body laterally (sideways). When you fire from a seated position, never rest your elbows on your knees. This transfers any movement of your leg directly to the sight.

c. **Firing**. Fully depress the safety before you try to squeeze the trigger. Then, squeeze—do not pull—the trigger. Since the Dragon has little recoil, many gunners move more when they pull the trigger than when they experience the missile's launch effects (recoil and backblast).

## 5-7. BREATH CONTROL

Aiming the Dragon is similar to aiming a rifle, except that the cross hairs must be kept on the desired impact point for 1 to 12 seconds following missile launch, depending on range to the target. Training must stress the importance of starting to hold the breath two seconds before squeezing the trigger and continuing to hold it while acquiring the target, firing, and tracking. That is, to prevent breathing from interfering with tracking, the gunner takes a breath and holds it while pressing the trigger. He must not breathe while tracking a target, because body movements cause the launcher to move. To check for breath control, the instructor watches the gunner's back.

## 5-8. TRACKING EXERCISES

Reacting properly to temporary obscuration is an important gunner skill. Occasionally, the target may be obscured by launch gases, dust, and so forth. The gunner's instinctive reaction is to look for a target. This can either cause erratic missile flight, or it can terminate the flight. Training prepares the gunner to "freeze" on a stationary target or to continue tracking at the established rate on a moving target until the target reappears. Instructors should simulate obscuration during DGT and DFTT exercises. They should show soldiers how an improper reaction causes the LOS to move outside the established aiming error limits. Most training should be conducted at moving targets to promote gunner concentration. A gunner who can hit a moving target consistently can hit a stationary target easily. Most *misses* occur on moving targets. Training in the tracking of a moving target should begin when a gunner achieves proficiency in assuming positions, sighting and aiming, and breathing.

## 5-9. RANGE PROCEDURES

Leaders normally consolidate qualification, verification, and sustainment range firing for gunners and assistant gunners at battalion or higher.

a. **Officer or NCOIC**. Table 5-3 (page 5-6) shows officer or NCOIC duties. (See also Appendix F.)

- Organizes the range.
- Assigns, coordinates, and supervises the firing lines.
- Issues fire commands and general instructions to the firing line.
- During all firing exercises, enforces safety precautions as prescribed in AR 385-62, local SOPs, and applicable range regulations.
- Stresses precision, a steady tracking rate, point training, and use of a firm firing position posture in all instructional firing.
- Fires exercises in the order listed in the firing tables. Ensures that instructors control the exercises using appropriate fire commands.
- Ensures that a qualified instructor inspects all DGTs, DFTTs, and sights before and after each firing day for cleanliness, serviceability, and operation.
- Instructs gunners and assistant gunners on duty assignments and range operating procedures before training them with the equipment. Divides them into teams and assigns each of them a position.

**Table 5-3. Officer or NCOIC duties.**

b. **Qualification Training**. For qualification, the gunner must fire the DGT.

c. **Verification Training**. A gunner must verify quarterly to meet qualification standards.

d. **Sustainment Training**. For monthly sustainment training, see Table 5-4 (page 5-7). The commander can select any target speed and monitor *DFTT* operation for gunner proficiency.

e. **Coach and DFTT Operator**. During instructional firing, a coach or DFTT operator is at each DFTT to instruct and assist the gunner. The coach or DFTT operator—

- Requires each gunner to observe safety precautions.
- Supervises each gunner's actions at the DFTT.
- Ensures gunners execute all the commands.
- Repeats orders and instructions to ensure understanding and timely execution.

- Reports misfires, malfunctions, and discrepancies to the OIC or NCOIC.
- Critiques the tracking runs.
- Sets target size on the computer.
- Sets obscuration on the computer.

| VEHICLE SPEEDS | | NUMBER OF ROUNDS | RANGE |
|---|---|---|---|
| KPH | MPH | | |
| 0 | 0 | 2 | 1,000 |
| 0 | 0 | 2 | 500 |
| 0 | 0 | 1 | 300 |
| 10 | 6 | 5 | 1,000 |
| 18 | 11 | 5 | 500 |
| 29 | 18 | 5 | 300 |

**Table 5-4. Sustainment speeds.**

f. **Training Team**. (Figure 5-5).

- DGT or DFTT gunner.
- Loader.
- Scorekeeper (optional).
- Safety monitor.
- One target vehicle operator, who does not participate in training.

**NOTE:** Rotate duty assignments between students.

**Figure 5-5. Training team.**

g. **Safety Precautions**. Implementing the precautions shown in Figure 5-6 ensures safety for all personnel.

- Observe the DFTT backblast area.
- Wear properly fitted earplugs.
- Reset the dummy weight before loading or reloading.

**Figure 5-6. Safety precautions.**

h. **Duty Assignments and Tasks of Training Team Personnel**. Figure 5-7 (page 5-8) shows training team personnel duty assignments and tasks.

| Duty Assignment | Tasks |
|---|---|
| **DFTT Gunner** | ☐ Avoids pointing either end of the DFTT at personnel. |
| | ☐ Fires only when cleared by the loader. |
| | ☐ Does not look at the sun or at searchlights when sighting. |
| | ☐ Considers all misfires to be hangfires and proceeds accordingly. |
| | ☐ Does not leave a loaded DFTT unattended. |
| | ☐ Prepares the DFTT for operation. |
| | ☐ Fires the DFTT. |
| | ☐ Reviews each tracking run with the DFTT operator. |
| **Loader** | ☐ Considers all misfires to be hangfires and proceeds accordingly. |
| | ☐ Helps perform misfire procedures. |
| | ☐ Loads or reloads the DFTT. |
| | ☐ Has all soldiers move their hands clear of the launcher before giving permission to fire. |
| | ☐ Informs the gunner when to fire. |
| **Scorekeeper** | ☐ Prepares the scorecard. |
| | ☐ Informs the instructor of the gunner's progress. |
| **Safety Monitor** | ☐ Ensures all participating team members use earplugs during the firing exercises. |
| | ☐ Keeps personnel clear of the DFTT's backblast area during the firing exercises. |
| | ☐ Reports all safety violations to the instructor for corrective action. |
| **Target Vehicle Operator** | ☐ Complies with all safety precaution warning signs. |
| | ☐ Drives the target vehicle. |

EXAMPLE

**Figure 5-7. Example format for checklist of duty assignments and tasks.**

## DANGER

THE DFTT USES HIGH VOLTAGES AND HIGH-EXPLOSIVE BLAST SIMULATORS.

YOU MUST FOLLOW ALL SAFETY PRECAUTIONS WHEN USING THIS EQUIPMENT. FAILURE TO DO SO COULD CAUSE YOU OR SOMEONE ELSE TO LOSE YOUR HEARING, SUFFER TRAUMATIC INJURY, OR EVEN DIE.

FOR INFORMATION ABOUT ARTIFICIAL RESPIRATION, SEE FM 21-11.

1. *HIGH VOLTAGES.*
   - AVOID TOUCHING THE METAL PINS WHEN CONNECTING OR DISCONNECTING POWER CORDS OR SYSTEM CABLES.
   - NEVER TRY TO DISASSEMBLE THE COMPUTER.
   - NEVER USE THE EQUIPMENT IF THE CATHODE RAY TUBE (CRT) OR CABLE IS DAMAGED.
   - FOLLOW THESE SAFETY PRECAUTIONS, OR YOU COULD DIE.
2. *HIGH-EXPLOSIVE BLAST SIMULATORS.*
   - WITHIN 70 METERS OF A LOADED BLAST SIMULATOR, WEAR HEARING PROTECTION.
   - HANDLE BLAST SIMULATORS CAREFULLY. IF POSSIBLE, WEAR FULL HELMET AND GLOVES.
   - KEEP BLAST SIMULATORS AWAY FROM FIRE.
   - NEVER HANDLE A DAMAGED BLAST SIMULATOR.
   - STAY CLEAR OF THE DANGER ZONE AT ALL TIMES.

CHAPTER 6

# GUNNER QUALIFICATION PROGRAM

*This chapter guides the unit through training soldiers to qualify as Dragon gunners. AR 614-200, paragraph 3-13, authorizes unit commanders to award the additional skill identifier (C2) to soldiers who successfully complete the qualification program. This chapter includes the qualification requirements and provides an example POI. (See Appendix G for gunner's performance test.)*

## 6-1. REQUIREMENTS

The gunner qualification program consists of three categories of instruction: informational classes (four of these), Dragon tasks (eight of these), and tracking qualifications.

a. **Informational Classes**.

- Introduction to the DGT.
- Introduction to the DFTT.
- Emergency decontamination procedures for the Dragon (Appendix H).
- Emergency destruction procedures for the Dragon (Appendix H).

b. **Tasks**. Soldiers must obtain a GO on each of the following tasks (the order is not important):

071-052-0001 — Maintain an M47 medium antitank weapon.
071-052-0003 — Construct a fighting position for an M47 medium antitank weapon.
071-052-0004 — Restore an M47 medium antitank weapon to carrying configuration.
071-052-0005 — Operate a night vision sight AN/TAS-5.
071-052-0006 — Engage targets with an M47 medium antitank weapon.
071-317-0000 — Prepare an antiarmor range card.
071-317-3302 — Prepare an M47 medium antitank weapon for firing.
071-317-3306 — Perform misfire procedures on an M47 medium antitank weapon.
071-317-3324 — Select a fighting position for an M47 medium antitank weapon.

c. **Tracking Qualification**. Soldiers must successfully engage 16 out of 20 targets on the DGT. Using the DGT requires an indoor training area.

## 6-2. PROGRAM OF INSTRUCTION

*Initial gunnery trainers* at Fort Benning, GA use the example POI shown in Figure 6-1 (page 6-2) to qualify *initial entry soldiers* on the Dragon. Trainers in each unit can adapt this POI to the resources available.

DEPARTMENT OF THE ARMY

BRAVO COMPANY 2ND BATTALION 29TH INFANTRY REGIMENT

FORT BENNING, GEORGIA 31905

ATSH-INB-B 9 AUG 99

MEMORANDUM FOR RECORD

SUBJECT: FIVE-DAY POI for Dragon Gunnery

1. Schedule. The following is a five-day POI training schedule for the Dragon Gunnery Course. It includes uniform, location and times for each class:

| DAY 1–LEE FIELD | SUBJECT | UNIFORM |
|---|---|---|
| 0900-1000 | Introduction to the M47 Dragon (Characteristics, Safety, and Limitations) | C |
| 1000-1100 | Prepare the M47 for Firing | B |
| 1100-1200 | Prepare and Restore the M47 for Firing | B |
| 1200-1300 | Eat Lunch | C |
| 1300-1330 | Demonstrate Correct M47 Firing Position | B |
| 1330-1430 | Demonstrate Target Engagement Techniques | B |
| 1430-1530 | Perform Misfire Procedures on the M47 Dragon | B |
| DAY 2–BLDG 4, RM 26 | SUBJECT | |
| 0900-1100 | Identify Combat Vehicles | C |
| 1100-1200 | Prepare an Antiarmor Range Card | C |
| 1200-1300 | Eat Lunch | C |
| 1300-1600 | DGT Firing Tables 1-4 | B |
| DAY 3–LEE FIELD | SUBJECT | |
| 0900-1000 | Review/Test Prepare an Antiarmor Range Card | C |
| 1000-1100 | Identify the six major components of an M47 Fighting Position | B |
| 1100-1200 | Maintain the M47 Dragon | B |
| 1200-1300 | Lunch | C |
| 1300-1400 | Operate the AN/TAS-5 nightsight | B |
| 1400-1600 | DFTT daysight firing. Table 5 (10 rounds) | A |
| 1600-1700 | Completion of DFTT nightsight firing during daylight conditions. Table 6 (5 rounds) | A |

Figure 6-1. Suggested POI for Dragon gunnery qualification.

| DAY 3—LEE FIELD | SUBJECT | UNIFORM |
|---|---|---|
| 1700-1800 | Dinner | C |
| 1800-1830 | Safety Briefing | A |
| 1830-UTC | Conduct DFTT Night Sight firing during darkness. Table 6 (5 rounds) | A |
| **DAY 4—BLDG 4, RM 26** | **SUBJECT** | |
| 0900-1000 | DGT Table 7, Practice Qualification | B |
| 1000-1200 | DGT Table 8, Qualification | B |
| 1200-1300 | Lunch | C |
| 1300-1400 | Completion of Tables 7 and 8 | B |
| 1400-UTC | Retest of Table 8 | B |
| **DAY 5—BLDG 4, RM 26/ COOLIDGE RANGE** | **SUBJECT** | |
| 0900-1030 | Maintenance | C |
| 1100-1130 | End of Course Critique | C |
| 1130-1230 | Lunch | C |
| 1230-1300 | Movement to Coolidge Range | B |
| 1300-UTC | Dragon Live Fire | B |

2. **Uniform code:**

A = LCE and BDU Cap

B = LCE and helmet

C = BDUs

**NOTE:** Instructors and students will wear the same uniform.

3. **Questions:** If you have questions about this memorandum, contact the undersigned.

PHIL S. CLIFFORD
SSG, USA
Master Trainer
CML 706-545-9918
DSN 835-9918

Figure 6-1. Suggested POI for Dragon gunnery qualification (continued).

CHAPTER 7

# SUSTAINMENT TRAINING PROGRAM

*This chapter recommends a sustainment program for units to conduct for qualified Dragon gunners. Training should progress into collective training for applicable MTPs and field manuals. Unit commanders can get qualified Dragon gunners either from the United States Army Infantry School or from a unit-run gunner qualification program (Chapter 6). After completing either of these training programs, each gunner must take part in sustainment training to maintain his tracking skills.*

## 7-1. GUNNER SUSTAINMENT TRAINING

Units conduct gunner sustainment training to maintain a high level of gunner proficiency. Training is divided into monthly, quarterly, and annual training phases.

a. **Monthly Sustainment Training**. Units conduct monthly training eight months of the year. The other four months, they conduct quarterly qualification or verification (on the DGT).

- One month, gunners perform half the tasks in Table 7-1 (page 7-2) using Firing Tables 1 and 2.
- In alternate months, gunners perform the other half of the tasks using Firing Tables 3 and 4.
- Each month, each gunner must score 16 out of 20 hits.

b. **Quarterly Training**. Each quarter, gunners fire practice qualification (Firing Table 7) and qualification (Firing Table 8). Each table includes 20 target engagements. Gunners must complete successfully all the tasks shown in Table 7-1. Anyone who fails to verify his qualification must retrain and retest. The qualification standards follow:

| | | |
|---|---|---|
| 19 or 20 hits | — | Expert |
| 17 or 18 hits | — | Sharpshooter |
| 16 hits | — | Marksman |

c. **Annual Training**. The commander follows one quarterly training period a year with an annual live-fire exercise. This exercise really tests the gunners' training. Commanders must ensure that only qualified Dragon gunners fire the live missiles. They must exclude from the live-fire exercise any gunner who fails to qualify on DFTT Table 6 or DGT Table 8. Allowing unqualified gunners to fire jeopardizes everyone else present.

## 7-2. COLLECTIVE TRAINING

Though the Dragon MILES offers an excellent way for gunners to keep their tracking skills honed (Table 7-2, page 7-2), leaders must still provide sustainment training.

| TASK NO | TASK | TNG FREQ | TIME | REMARKS |
|---|---|---|---|---|
| — | Explain Emergency Decontamination Procedures for an M47 Medium Antitank Weapon | M | 10 min | Oral Presentation |
| — | Explain Emergency Destruction Procedures for an M47 Medium Antitank Weapon | M | 10 min | Oral Presentation |
| — | Tracking Sustainment Training | M | 4 hrs | Practical Exercise |
| 071-052-0001 | Maintain an M47 Medium Antitank Weapon | $M^1$ | 10 min | Practical Exercise |
| 071-052-0004 | Restore an M47 Medium Antitank Weapon to Carrying Configuration | $M^1$ | 5 min | Practical Exercise |
| 071-317-3306 | Perform Misfire Procedures on an M47 Medium Antitank Weapon | $M^1$ | 15 min | Practical Exercise |
| GTA 17-2-11<br>GTA 17-2-13 | Recognize Friendly and Threat Armored Vehicles | $M^2$ | 15 min | Practical Exercise |
| 071-052-0003 | Construct a Fighting Position for an M47 Medium Antitank Weapon | $M^2$ | — | Perform as part of an ARTEP[3] or FTX |
| 071-317-3302 | Prepare an M47 Medium Antitank Weapon for Firing | $M^2$ | 30 sec | Practical Exercise |
| 071-317-3324 | Prepare an Antiarmor Range Card | $M^2$ | 15 min | Practical Exercise |
| 071-052-0006 | Engage Targets with an M47 Medium Antitank Weapon | A | 8 to 16 hrs | Practical Exercise |

M = Every month
$M^1$ = First month
$M^2$ = Second month
A = Annually

3 ARTEPs are available at Reimer Digital Library.

**Table 7-1. Training tasks for Dragon gunners and assistant gunners.**

| TRAINING SEQUENCE | RECOMMENDED FREQUENCY: TIMES PER YEAR | EQUIPMENT USED |
|---|---|---|
| Squad, platoon, or company FTX | 4 | Dragon/ MILES |
| Battalion FTX | 2 | Dragon/ MILES |
| Platoon or company MOUT exercise | 1 | Dragon/ MILES |
| Squad, platoon LFX | 2 | Dragon/ MILES or LTIDs (laser-target interface devices) |
| Externally evaluated battalion MTP | 1 | Dragon/ MILES |

**Table 7-2. MILES training program.**

CHAPTER 8

# EMPLOYMENT

*This chapter discusses roles and duties, employment considerations, and target-engagement techniques for employing the Dragon in defensive operations. Though this chapter discusses the Dragon as an infantry weapon, the techniques described apply to any situation in which when a soldier uses a Dragon, regardless of the type of unit. Specific discussions include selection and preparation of firing positions, target-engagement techniques, and fire-control procedures. To fully understand and properly integrate the Dragon into unit TTP, leaders must know* FM 7-7, FM 7-7J, FM 7-8, FM 7-10, *or* FM 71-1, *whichever applies.*

## Section I. EMPLOYMENT IN THE DEFENSE

This section provides guidance for employing the Dragon in defensive operations. This information applies to all types of infantry and other types of units. In infantry units, the platoon has the mission in the defense to repel the enemy's assault-by-fire and close combat. The Dragon has combat characteristics that are important to the defense. (FM 7-8, FM 7-7J, and FM 7-10 cover defensive techniques thoroughly.) The Dragon(s) can—

- Destroy or immobilize armored vehicles, depending on type.
- Deliver accurate fire, day or night.

### 8-1. FIRING POSITIONS

Figure 8-1 shows the three types of firing positions: primary, alternate, and supplementary.

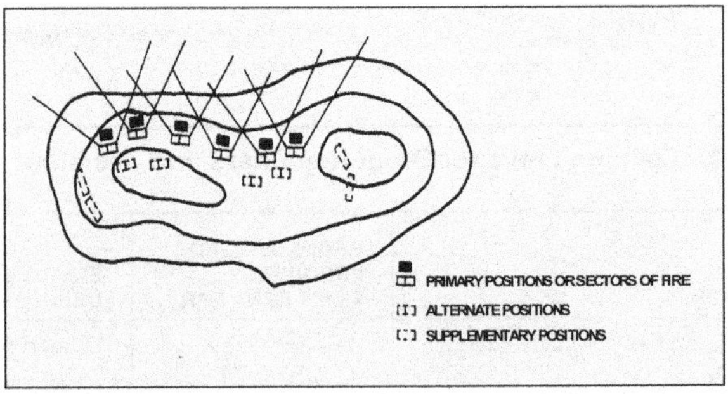


**Figure 8-1. Primary, alternate, and supplementary positions.**

a. **Primary Position.** From this position, the gunner or team can cover the assigned sector of fire. The position should offer good observation, good cover and concealment, and a good field of fire.

b. **Alternate Position.** The *alternate position* is either to the flank or slightly to the rear of the primary position. The gunner or team must be able to cover the same sector of fire as they could from the primary position. He (or they) occupies the alternate position when they must leave or cannot occupy the primary position.

c. **Supplementary Position**. The *supplementary position* covers avenues of approach and any TRPs not covered by the primary and alternate positions. This position usually falls close enough to the primary position to share mutual support with other positions.

## 8-2. MOVEMENT

Whether mounted or dismounted, the gunner should observe certain basics when moving to and from firing positions.

a. Avoid disturbing natural foliage.

b. Move in and around positions as little as possible.

c. Move into the position from the rear. In the defense, the only movement forward of the position should be the gunner and the other soldiers clearing the fields of fire. The leader checks for cover and concealment and paces the distances for the range card.

d. Select good covered and concealed routes to and from positions. Gullies and reverse slopes are excellent options for protection and ease of movement.

## 8-3. SELECTION OF FIGHTING POSITIONS

The two main factors in positioning the Dragon include gunner protection and effective use of the weapon's capabilities. The gunner must remain exposed while tracking his targets. When firing at a target at maximum range, he remains susceptible to counterfires for as long as 12 seconds. The gunner avoids positions that would force him to fire into the sun, which could affect his ability to track the target. Many of the steps for protecting gunners also make the most of the Dragon's capabilities. Gunners can enhance mission accomplishment by following some basic rules when selecting positions:

a. Use natural cover and concealment. Choose positions where the terrain provides *cover* from enemy fire and *concealment* from enemy ground and aerial observation (Figure 8-2).

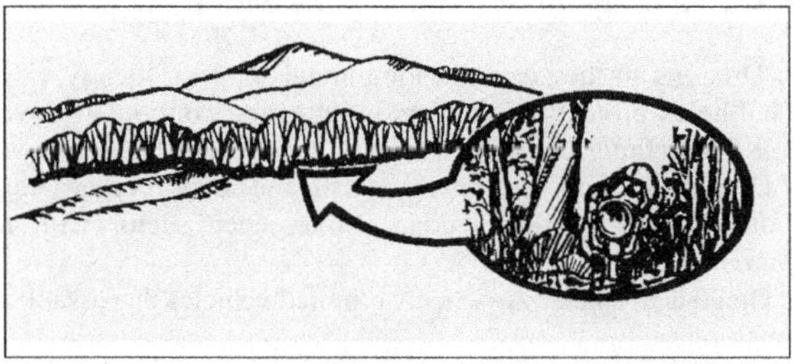

**Figure 8-2. Use of natural cover and concealment.**

b. Try to choose positions where the gunner can engage the enemy's flank or rear from behind frontal cover. A gunner firing obliquely should do so from beneath the protection of a parapet or natural cover. Dragon missiles probably will not defeat a tank hit in the frontal 60-degree arc. You will also find other armored vehicles easier to kill from the flanks and rear. The enemy has a tougher time tracing the origin of a Dragon round fired from his flank than one fired head-on.

(1) From an oblique weapon position, a gunner can provide frontal cover to protect against direct suppressive fires. An oblique position also helps to conceal his location from the view of any person or vehicle approaching from the front (A, Figure 8-3).

(2) In the attack, a tank orients firepower and observation mainly to the front. This helps keep the tank from detecting the launch site (B, Figure 8-3).

c. Avoid using Dragon positions that must engage targets from the front (especially tank targets).

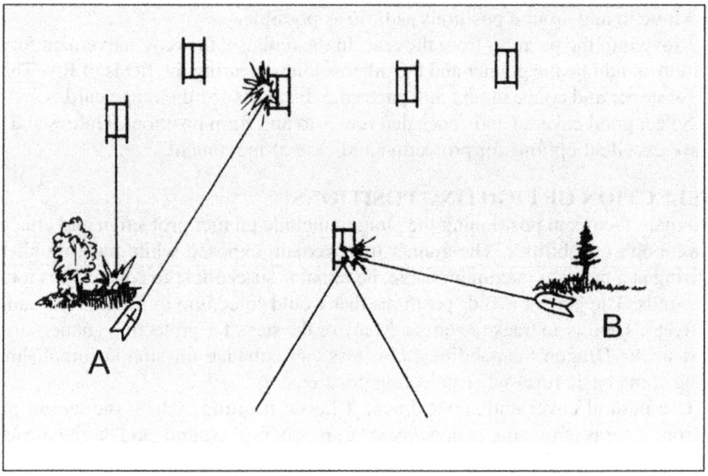


**Figure 8-3. Engagement of the enemy with flank or rear shots.**

d. Position Dragons so they can provide mutual support. Ideally, position them 300 meters apart. This helps protect the gunners by ensuring continuous coverage of enemy armored vehicles. To do this—

(1) Employ Dragons so their fires interlock with and support other Dragons, TOWs, or tanks. Ensure that sectors of fire overlap. Cover each sector with more than one antiarmor weapon.

(2) Position Dragons to engage any enemy armored vehicles that assault another Dragon, TOW, or tank position.

e. For security, integrate Dragon gunners with nearby infantry. Leaders should provide local security for any Dragon gunner employed away from the squad or platoon such as a Dragon team or armor-killer team.

## 8-4. PREPARATION OF FIGHTING POSITIONS

The enemy must not learn the true location of the Dragon. Clearing away loose debris behind the launcher, wetting down the backblast area, and covering the ground with shelter halves reduces the Dragon's launch signature (backblast). To prevent detection, soldiers move only in and around the position. As long it can see the target clearly, the unit can use indirect fires (HE, smoke, and WP) and small-arms weapons to distract the enemy. Other deception measures include preparing partly visible dummy positions to draw enemy fire away from the actual positions, then positioning Dragons on less obvious firing positions. The Dragon fighting position offer have unobstructed fields of fire, mask clearance (minimum dead space that could hide targets in the sector), and a clear backblast area. Just as it can do with other weapons organic to the platoon, the unit can employ the Dragon either from hasty or improved positions. Soldiers situate and orient a fighting position to cover a sector of fire.

a. After receiving a sector of fire and firing location from the leader, the gunner constructs the Dragon position to cover the sector. He clears only what he must clear for effective fields of fire. He camouflages the position using available materials and improves the position as time permits.

b. Leaders must consider the backblast and the muzzle blast when employing the weapon. To prepare a Dragon fighting position, the gunner follows these guidelines:

(1) When the gunner fires from an improved position, the muzzle end of the launcher must extend 15 centimeters (6 inches) beyond the front of the hole. The rear of the launcher must extend out over the back of the hole. As the missile leaves the launcher, the unfolding stabilizing fins require at least 15 centimeters (6 inches) of clearance above ground. The position should offer protection to the front (a parapet) or other natural or man-made cover.

(2) Gunners clear the ground in front of and behind the position. They remove rocks, sand, and debris. This prevents a dust cloud from forming when the gunner fires. Dust would obscures a gunner's vision and marks the location for enemy observers. When the gunner must fire in only one direction, a one-person fighting position works best (Figure 8-4, following; and Figure 8-5, page 8-5).

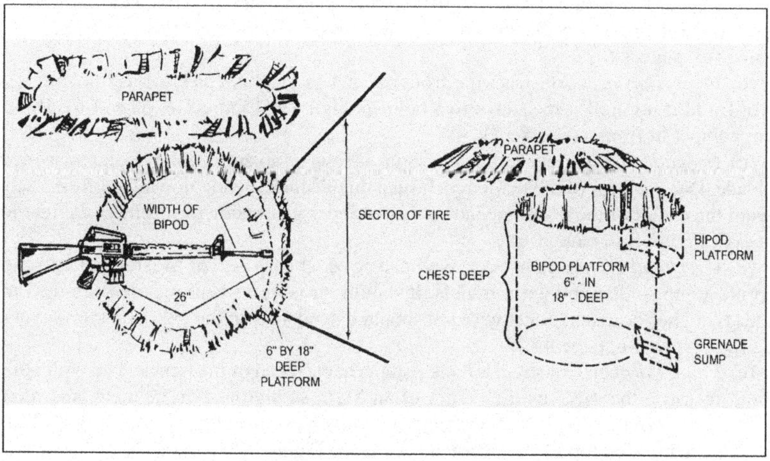


**Figure 8-4. Construction of a one-person fighting position.**

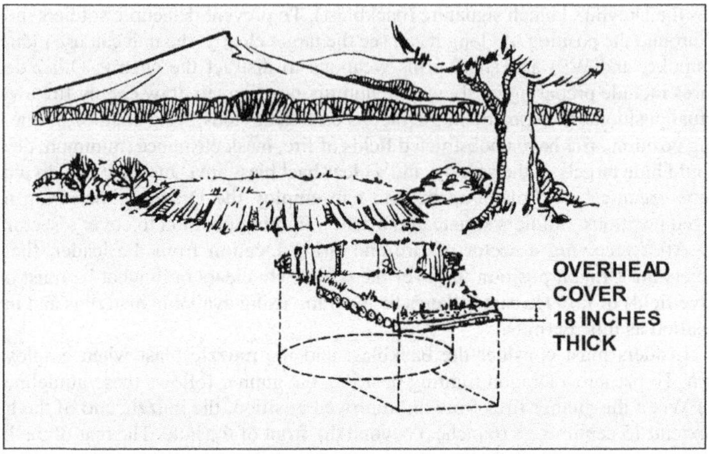


**Figure 8-5. Construction of overhead cover.**

(3) The gunner should be positioned to fire obliquely. This protects the gunner from frontal fire while he engages the target from the flank. If necessary, the gunner can fire to the front as well as to the oblique from a one-person fighting position.

(4) The wedge shape of the two-person fighting position gives the gunner frontal protection. It also allows the gunners to engage their targets from the oblique or flank. The team constructs the position as follows:

(a) *Trench*. Construct the trench position the length of three M16s, in an inverted "V." Dig it waist-deep. Make the trench waist-wide plus about 15 centimeters (6 inches) (A, Figure 8-6, page 8-6).

(b) *Front Parapet*. Construct the front parapet as long and as wide as the length of an M16. Build it up until it measures two helmets high (B, Figure 8-6, page 8-6). Build the front parapet in front of the trench.

(c) *Grenade Sump*. Make a grenade sump as long as an entrenching tool, and as wide as its blade. Dig the floor of the main trench such that it slopes gently downward from each end toward the center of the position, and so that it slopes gently downward from the rear to the front (C, Figure 8-6, page 8-6).

(d) *Overhead Cover*. Construct overhead cover at each end of the trench large enough to protect one soldier and extra rounds. It should measure 31 centimeters deep by 1 meter wide (12 inches deep and 3 feet wide). It should extend 46 centimeters (18 inches) over each side (D, Figure 8-6, page 8-6).

(e) *Flank Parapet*. Construct a flank parapet at each end of the trench. The width of each should measure the same as the length of an M16, as high as two helmets, and a length sufficient to provide good flank protection. To increase overhead protection, build flank parapets are built on top of the overhead cover (E, Figure 8-6, page 8-6).

(f) *Bipod Trench*. Dig a bipod trench for each sector of fire. The back of the bipod trench should measure 10 to 15 centimeters (4 to 6 inches) forward of the main trench. Make the bipod trench two helmets long, one helmet wide, and 15 centimeters (6 inches) deep (F, Figure 8-6).

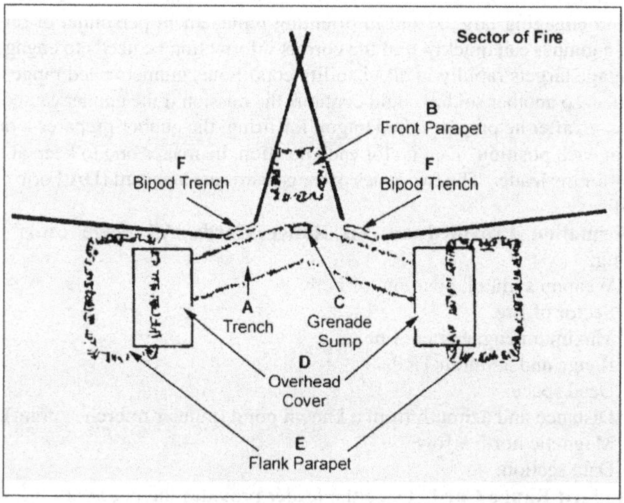


**Figure 8-6. Construction of a two-man fighting position.**

(g) *Front Cover*. Sometimes the gunner can fire only in one direction. If so, construct front cover so the gunner should engage targets from the flank (G, Figure 8-6). Also construct cover and concealment from other directions (Figure 8-7).

**Figure 8-7. Fire in one direction.**

## 8-5. PREPARATION OF RANGE CARDS

A range card consists of a sketch of the terrain that a specific weapon system covers. The range card contains information that helps in planning and controlling fires, in quickly detecting and engaging targets, and in orienting replacement personnel or units. Using a range card, a gunner can quickly find the correct information he needs to engage targets. In order to engage targets rapidly in all visibility conditions, gunners need range cards. They also need them so another soldier could continue the mission if the gunner can no longer fire. For this reason, after he prepares the Dragon for firing, the gunner prepares a range card in duplicate for each position. That is, for each position, he makes one to keep at the position and another for the leader. The two types of range cards are standard (DA Form 5517-R) and field-expedient.

a. **Information Provided on All Range Cards**. All range cards must show the following:

- Weapon symbol, position, or both.
- Sector of fire.
- Maximum engagement line.
- Range and azimuth TRPs.
- Dead space.
- Distance and azimuth from a known point (gunner reference point).
- Magnetic north arrow.
- Data section.

b. **Standard Range Card**. Once the leader provides the necessary information, the gunner prepares a standard range card in duplicate (Figure 8-8, page 8-8).

c. **Field-Expedient Range Card**. Because he may not be able to find any standard range cards in a combat situation, the gunner can draw one on anything available. He prepares a field-expedient range card the same as he would any other range card, except that he uses the weapon symbol to show only the location of the weapon system (Figure 8-9, page 8-9).

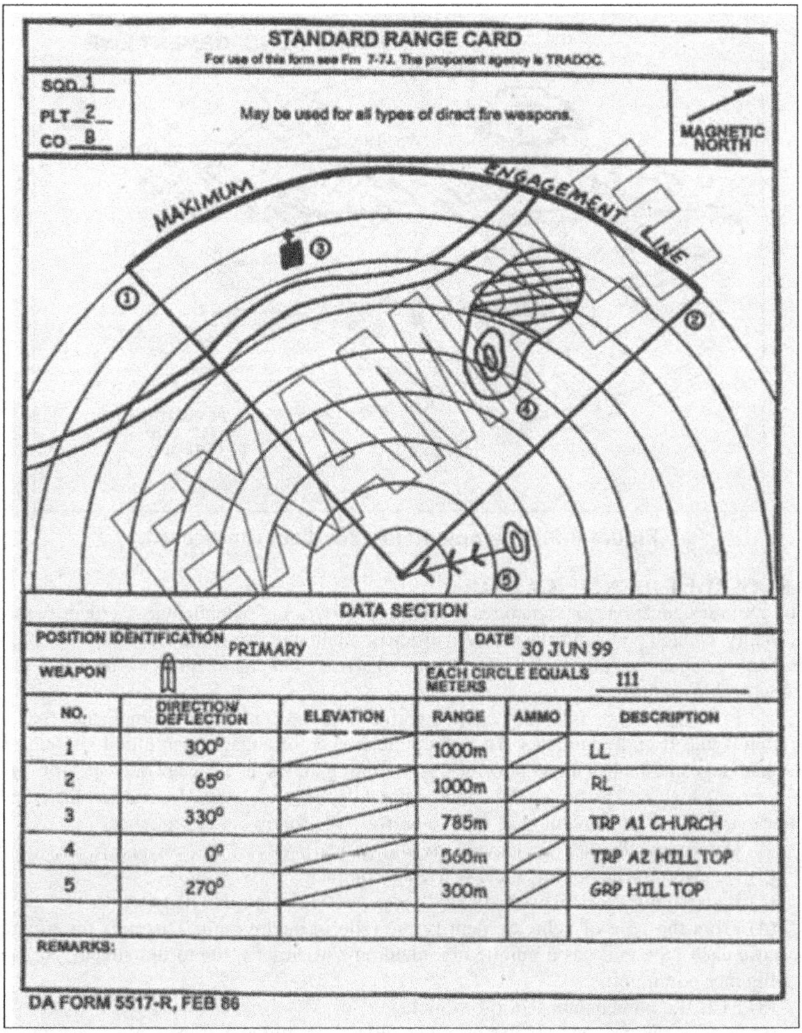

STANDARD RANGE CARD

For use of this form see FM 7-7J. The proponent agency is TRADOC.

SQD. 1

PLT. 2

CO. B

May be used for all types of direct fire weapons.



DATA SECTION

POSITION IDENTIFICATION PRIMARY

DATE 30 JUN 99

WEAPON

EACH CIRCLE EQUALS 111 METERS

| NO. | DIRECTION/ DEFLECTION | ELEVATION | RANGE | AMMO | DESCRIPTION |
|---|---|---|---|---|---|
| 1 | 300° | | 1000m | | LL |
| 2 | 65° | | 1000m | | RL |
| 3 | 330° | | 785m | | TRP A1 CHURCH |
| 4 | 0° | | 560m | | TRP A2 HILLTOP |
| 5 | 270° | | 300m | | GRP HILLTOP |
| | | | | | |

REMARKS:

DA FORM 5517-R, FEB 86

**Figure 8-8. Example completed DA Form 5517-R, Standard Range Card.**

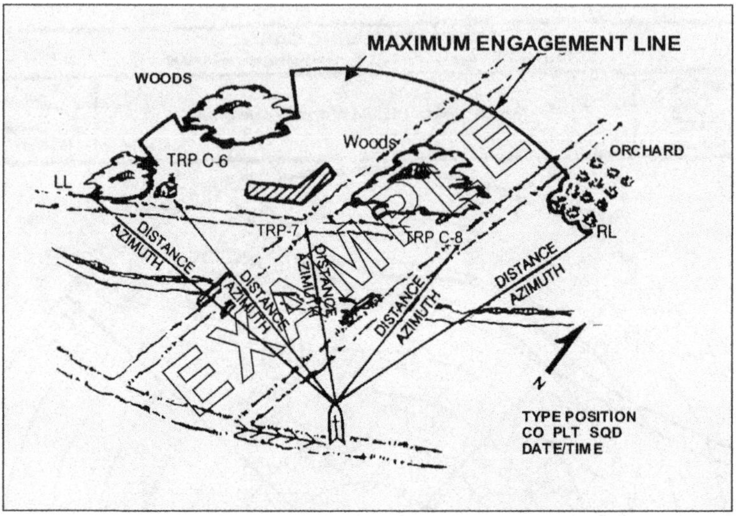


**Figure 8-9. Example field-expedient range card.**

## 8-6. TARGET IDENTIFICATION

The activities, locations, or signatures (visual or otherwise) of potential targets identify them as enemy. Dragon gunners must receive sufficient training to recognize the sizes, shapes, and thermal images of all types of targets. Turrets and main guns offer the most recognizable identifiers.

a. Friendly foreign units may operate with or near US forces. This complicates the task of identifying friendly vehicles. To reduce the chance of engaging an allied vehicle, the commander can establish target priorities. Antiarmor gunners then engage only specific *types* of enemy vehicles. Which type the commander tells them to engage depends on the enemy situation. Dragon gunners must be able to perform the following in sequence:

(1) Determine whether a vehicle is tracked or wheeled.

(2) Determine whether a vehicle is friendly or enemy.

(3) If an enemy vehicle, use Table 8-1, page 8-10 to determine its type.

(4) From the type of vehicle, identify the type of enemy unit. This aids the gunner, because each type unit has a unique organization and target value to the gunner, S2, and intelligence community.

(5) State the nomenclature of the vehicle.

b. Most weapons and vehicles produce telltale signatures. For example, most tracked vehicles use diesel fuel, which emits a large amount of black smoke. Tracked vehicles make more noise than wheeled vehicles. Antiarmor units can use these and other signatures to help them locate and identify enemy targets.

| RECON | APC | IFV | TANK |
|---|---|---|---|
| • AML-90<br>• BRDM-1<br>• BRDM-2<br>• EE-9<br>• M-93 Scorpion | • BTR-50<br>• BTR-60<br>• BTR-70<br>• BTR-80<br>• LAV-25<br>• M-113<br>• MT-LB<br>• PRC TYPE 63/YW 531<br>• V-150 | • BMP-1<br>• BMP-2<br>• BMP-3<br>• BRM<br>• BMD<br>• M2A2 Bradley<br>• VTT-323 (M1973) | • Centurion<br>• Challenger<br>• Chieftain<br>• M1A1 Abrams<br>• M60<br>• M1985 (NK)<br>• PT-76<br>• T-54<br>• T-55<br>• T-62<br>• T-64<br>• T-72<br>• T-80 |

**Table 8-1. Vehicles categorized by function.**

c. Dragon gunners mainly detect target signatures by sight, sound, and smell. If they detect anything suspicious or unusual, they should thoroughly investigate it. Sun shining off a flat surface, such as off a windshield, sounds of diesel or turbine engines, or the clanking or squeaking of end connectors can indicate the locations of targets.

(1) ***Soldier Signatures***.

- Fighting positions.
- Trash.
- Cut or missing vegetation (cleared for fields of fire or camouflage).
- Freshly dug earth (may indicate a fighting position).
- Noise from equipment or talking.
- Light from a match, cigarette, or fire.

(2) ***Tracked Vehicle Signatures***.

- Large dust clouds.
- Diesel smoke.
- Noise made by tracks and engine.
- Vehicle tracks on the ground.
- Distinctive silhouette or shape.

(3) ***Antitank Weapon Signatures***.

- "Swish" of missile launch.
- Long, thin wires in brush, trees, or along the ground.
- Dismounted soldier looking through a periscope-type device. (Launcher could be up to 100 meters away from him.)

(4) ***Aircraft Signatures***.

- Reflection of the sun from aircraft canopies and rotor blades.
- Vapor trails.

- Dust and movement of foliage caused by a hovering helicopter.
- Sound of a turbine engine (high-pitched whirring sound).

(5) ***Obstacles and Mines***.

- Loose dirt or dirt that has been disturbed in a regular pattern.
- Areas where large trees have been removed.

**8-7. TARGET RECOGNITION BY TYPE**

Learning to recognize targets by type presents little challenge. However, identifying them as friendly or enemy requires careful study and attention to detail. Tanks are most difficult, because many friendly and enemy tanks share many design similarities. When camouflaged and moving at a distance of 1,500 to 2,000 meters, the gunner may not detect the differences. Soldiers must know which friendly and threat armored vehicles most likely will appear on the battlefield (STP 21-1-SMCT, available in Reimer's Digital Library). Soldiers can use training aids, such as GTA 17-2-13, to study the armored vehicles of other nations (Figure 8-10, page 8-12). To identify most armored vehicles or tanks, the gunner considers the type, location, and absence or presence of certain equipment. Specifically, he looks at the suspension system, turret, and main gun. However, he should remember that, just like friendly forces, the enemy also uses camouflage and deception.

a. **Suspension System**. Vegetation and terrain often conceal a vehicle's suspension system, which makes it the least identifiable part of the vehicle. If the gunner can see it, he can distinguish a suspension system by its—

- Road wheels and support rollers.
- Road wheels only.
- Number of road wheels.
- Spacing between road wheels.
- Armored skirt.

b. **Turret**. The gunner distinguishes turrets by these characteristics—

- Position on the hull: well forward, center, or to the rear.
- Presence, absence, or location of a searchlight.
- Shape of the turret: rounded, elongated, or boxy.
- Externally mounted storage racks and other equipment.

c. **Main Gun**. The gunner distinguishes tank main guns by looking for—

- A bore evacuator on the gun tube.
- A muzzle brake or blast deflector.
- The presence or absence of a thermal jacket.

d. **Commander's Station (Some Tanks)**. Normally a simple hatch or cupola, the commander's station projects from the top left or right side of the commander's turret.

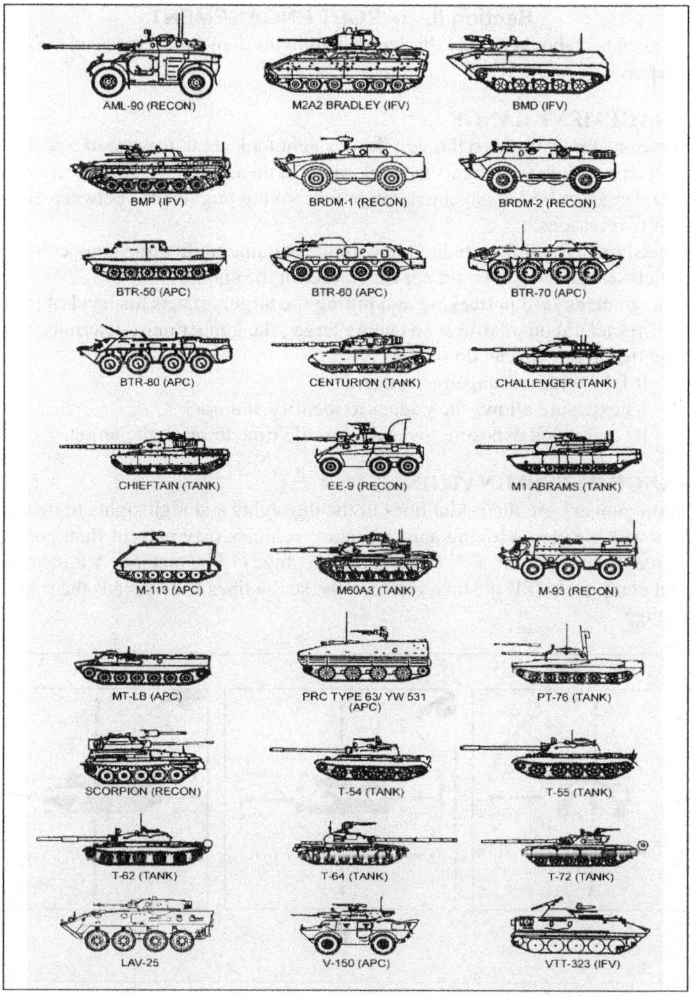


**Figure 8-10. Differences between armored vehicles.**

## 8-8. THERMAL IDENTIFICATION

Using the AN/TAS-5 to identify targets by their thermal signature presents a challenge. Doing it successfully requires extensive training. Appendix I discusses in detail how to identify targets by their thermal signatures. Gunners use the same steps to identify vehicles thermally as they would to identify them through the daysight.

## Section II. TARGET ENGAGEMENT

The Dragon gunner should engage the enemy within his own capabilities and the capabilities of the weapon.

### 8-9. ENGAGEMENT RANGE

At 1,000 meters, targets viewed through the daysight look about the size of a postage stamp. Engaging a target while viewing its thermal image in the nightsight presents more difficulty.

a. The Dragon's best engagement range for moving targets falls between 200 and 800 meters, for two reasons:

(1) The shorter flight time reduces the gunner's vulnerability to enemy counterfires.

(2) At closer ranges, the target appears larger in the sights.

b. The gunner's skill in tracking and hitting the target reflects his level of proficiency. Before he fires a Dragon missile at an enemy target, the gunner must determine whether he can engage the target. He can do so if—

- It falls within engagement range.
- Its exposure allows the gunner to identify and track it.
- Its continued exposure gives the missile time to reach the target.

### 8-10. RANGE DETERMINATION

The Dragon gunners use the stadia lines in the daysights and nightsights to determine if a target falls within range. Moving and stationary vehicles may present flank, oblique, and frontal or rear targets (Figure 8-11). At maximum range (1,000 meters), A 6-meter (20-foot) long target completely fills the area between the stadia lines and exceeds the stadia lines at a closer range.

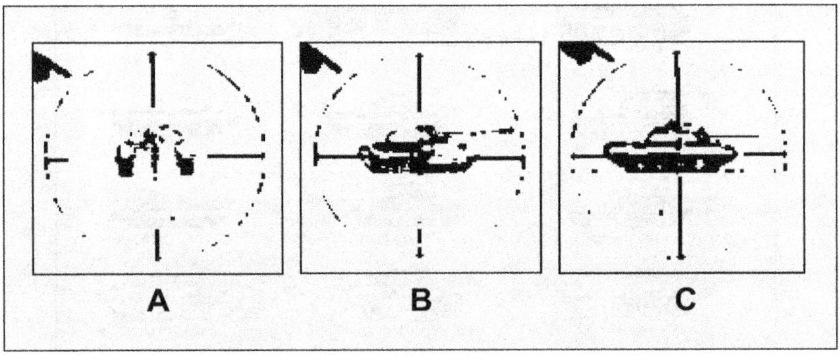


**Figure 8-11. Frontal, oblique, and flank targets.**

a. **Flanking Targets (Full Stadia).** Adjust the sight picture by moving the launcher so the target centers between the stadia lines (Figure 8-12, page 8-13).

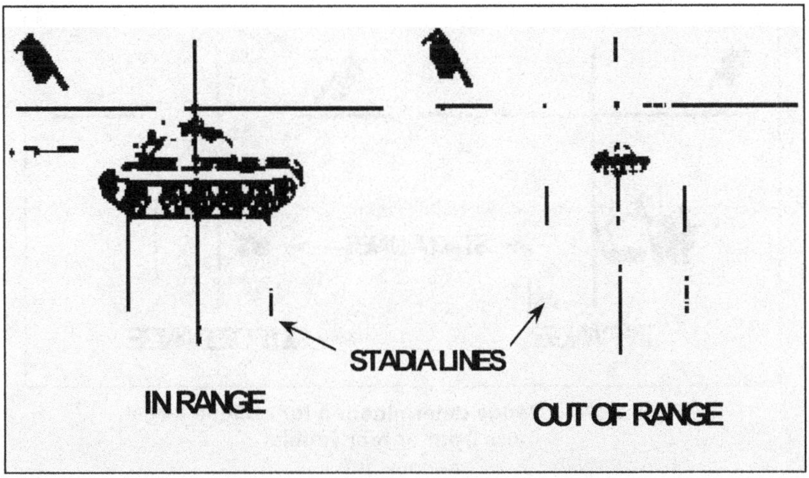


**Figure 8-12. Range determination for flank target.**

b. **Oblique Targets**. If you can see more of the flank, use the full-stadia method (Figure 8-13). If you see more of the front or rear, use the half-stadia method (Figure 8-14, page 8-15).

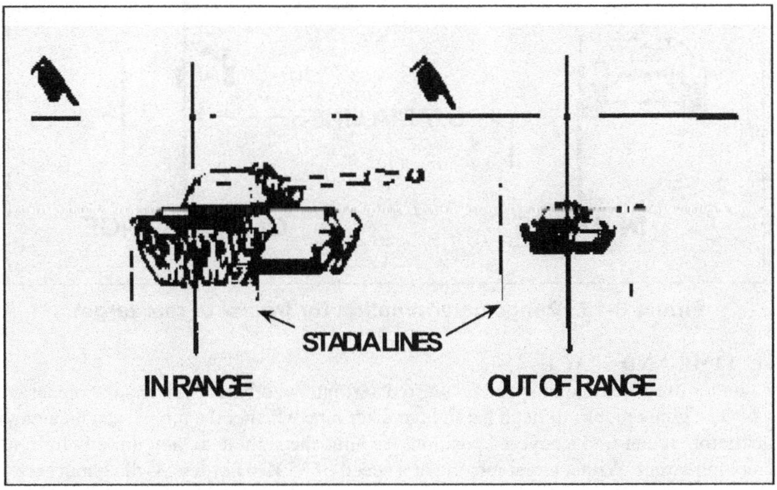


**Figure 8-13. Range determination for oblique target, more flank visible.**

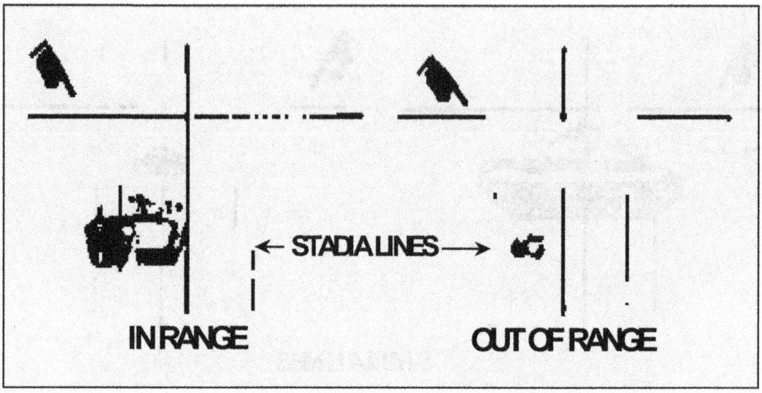


**Figure 8-14. Range determination for oblique target, more front or rear visible.**

c. **Frontal (Head-On) or Rear (Going Away, Half-Stadia) Targets**. Adjust the sight picture by moving the launcher to align the vertical cross hair and one of the stadia lines on the target (Figure 8-15).

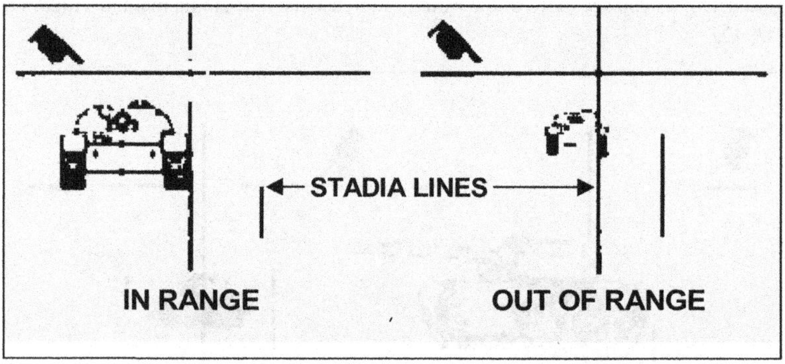


**Figure 8-15. Range determination for frontal or rear target.**

## 8-11. TIME AND SPACE

The gunner uses the time or space factor to determine whether he can engage the target.

a. The gunner looks through his sight to determine whether the missile can hit a moving target before it can find a covered position. He aims the sight at a point directly in front of the moving target. With a target moving at a speed of 35 KPH or less, if the gunner sees no obstructions or covered areas on the apparent path of the target, he can destroy the target (Figure 8-16, page 8-16).

Figure 8-16. Target able or unable to reach cover in time.

b. If the gunner places the cross hairs center of mass and fires at once, the target will not reach the protection of the hill in time (Figure 8-17). If the gunner does fire—the target will reach the protection of the hill before the missile hits (Figure 8-18).

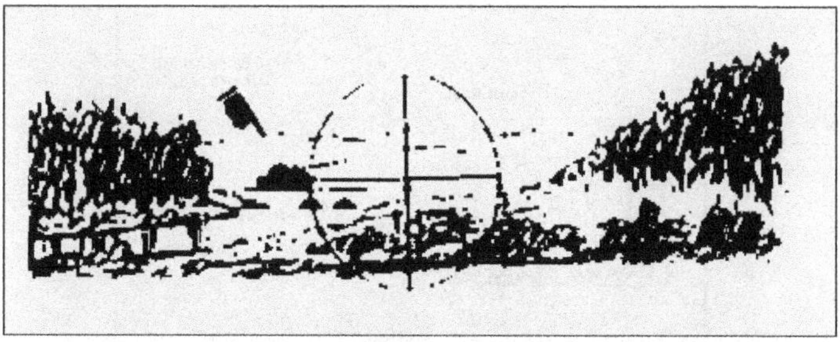

Figure 8-17. Target unable to reach cover in time.

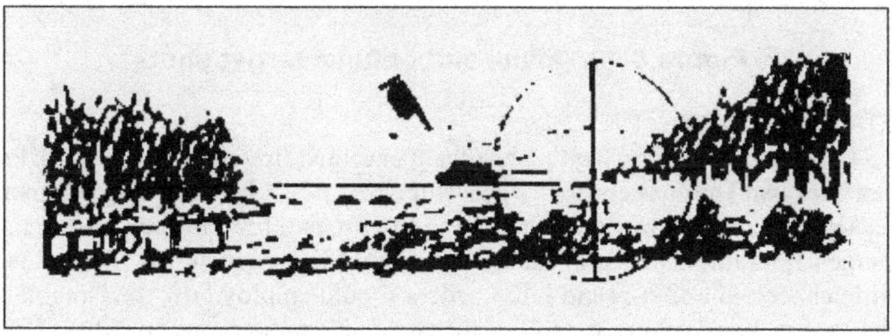

Figure 8-18. Target able to reach cover in time.

## 8-12. VULNERABLE POINTS OF ARMORED VEHICLES

After the gunner decides that he can engage a target, he should try to hit it at its weakest points. Leaders can help by positioning Dragons where the gunners can take advantage of those weak points.

a. To do so, the gunners must *know* the weak points of each type of enemy armored vehicle. An enemy armored vehicle usually has the most armor protection on its front glacis (slope). It has less on its flanks, and still less on its top, back, and belly. Any armored vehicle's weakest areas include its internal fuel tanks, ammunition storage areas, and engine. Destroying the engine not only immobilizes the vehicle, but may ignite ruptured fuel lines, causing a fire or explosion.

b. Gunners have a better chance to get a kill with oblique or flanking shots. They must keep themselves ready for opportunities. For example, a potential target may expose its flank when it tries to bypass an obstacle or evade an oncoming ATGM. Gunners produce more mobility kills with flank and oblique shots. However, they can do so even if they only hit the vehicle's wheels, track, or suspension system (Figure 8-19). Flank shots deflect off armor less often than oblique shots. The top or bottom (belly) of an enemy armored vehicle may show briefly, while the vehicle breaches an obstacle or antitank ditch, fords a river with steep banks, or traverses a shallow valley. By carefully analyzing the terrain in the assigned sector of fire, the gunner can determine where approaching armor units will expose their weaker armor.

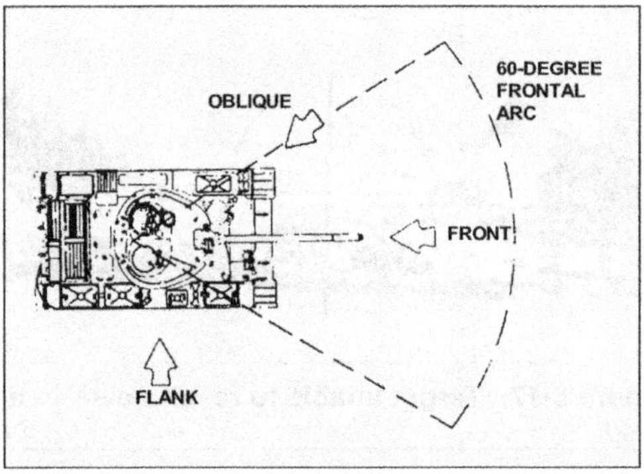


**Figure 8-19. Flank and oblique target shots.**

## 8-13. FIRE CONTROL

Effective Dragon fire requires well-planned and executed fire-control measures. Proper fire control ensures that Dragons engage targets at the best possible times. Leaders should never endanger Dragon gunners by ordering them to fire prematurely. Firing at a target traveling out of range offers little chance of success. So would firing when, for any reason, a target offers little chance of a first-round kill. Leaders should employ only the Dragon best sited to hit the target. Lack of fire control reduces a unit's antiarmor capability. Under ideal conditions, Dragons dispersed as far apart as 1,600 to 2,000 meters can concentrate their fires on the same group of targets (Figure 8-20).

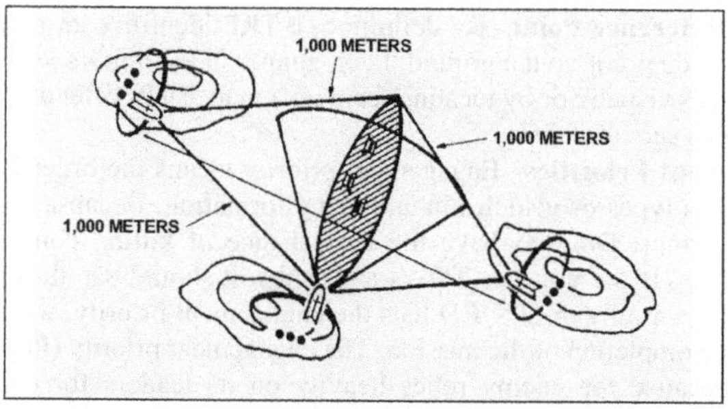


Figure 8-20. Dragon dispersion.

a. **Fire Control Methods**. Effective fire control methods prevent the wasteful firing of more than one Dragon missile at the same target, and they prevent premature firings. Unnecessarily firing a Dragon may disclose the location of the defensive position to the enemy. Dragon fire control methods include sectors of fire, TRPs, engagement priorities, fire patterns, and fire commands.

b. **Sector of Fire**. A sector of fire refers to an area limited by boundaries and assigned to a unit or weapon to cover by fire (Figure 8-21). The gunner or unit assigned a sector of fire may fire only within that sector. Leaders make sure that sectors overlap in order to cover all areas and so that the Dragon gunners do not hesitate when enemy vehicles come into range.

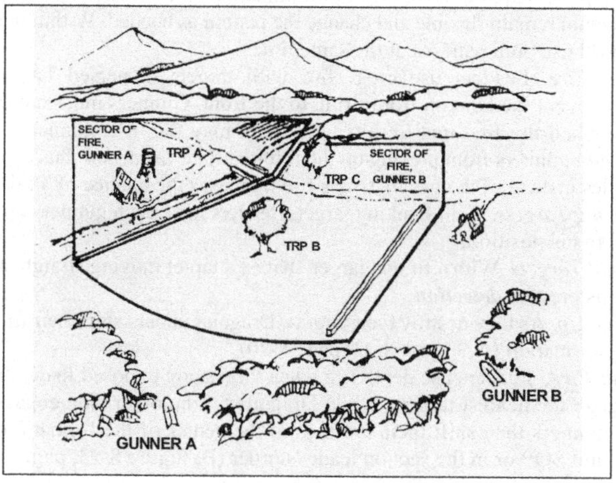


Figure 8-21. Sector of fire.

c. **Target Reference Point**. By definition, a TRP identifies an easily recognizable, natural or man-made point on the ground. Each gunner in the unit's sector must know how to identify all TRPs visually or by locating them on a map. TRPs offer a great way to define visually a gunner's sector of fire.

d. **Engagement Priorities**. Engagement priority means the order in which Dragons engage the various types of vehicles in an enemy formation. Because many enemy tanks have improved armor, Dragons have the best chance of killing command and control vehicles and APCs. If the unit has TOWs and tanks, it should use those weapons to kill enemy tanks. The warning or OPORD lists the engagement priority, which includes as its main priority the completion of the mission. The engagement priority (Figure 8-22) lists C2 vehicles first, because the enemy relies heavily on its leaders for combat operations. Eliminating leadership will cause confusion and disrupt the enemy's mission.

1. Command and control vehicles.
2. Armored personnel carriers.
3. Antiaircraft vehicles.

**Figure 8-22. Example of engagement priority.**

e. **Fire Patterns**. Fire patterns describe the relationship between Dragons and their targets. Firing patterns help leaders control gunners. They also give gunners specific targets to focus on. Two basic patterns include cross fire and depth fire. Leaders should not fixate on one, but should remain flexible and change the pattern as needed. Within the unit's sector, the leader could use both patterns at the same time.

(1) ***Cross Fire***. Leaders use cross fire with targets dispersed laterally or when obstructions prevent the Dragon from firing to the front. Gunners must stagger their cross fire. That is, when the first missile hits, only then may the next gunner fire. Cross fire prevents Dragon gunners from picking up the infrared flare from, and thus controlling, each other's missiles instead of their own. Cross fire increases the chance of a kill.

(a) *Flanking Targets*. With flanking targets, leaders have each gunner engage the target at a diagonal to his position.

(b) *Frontal Targets*. With a frontal target, that is, a target moving straight at the Dragon, cross fire helps prevent ***detection***.

(c) *Follow-Up*. As they destroy their targets, Dragon gunners shift their fire to the center of the enemy formation (A, Figure 8-23, page 8-20).

(2) ***Depth Fire***. Leaders use depth fire when targets are exposed in depth. Dragons on one side engage the nearest targets, while Dragons on the other side engage the farthest targets. The gunners then shift their fire toward the center of the formation. This can be specified by unit SOP or in the section leader's order (B, Figure 8-23, page 8-20).

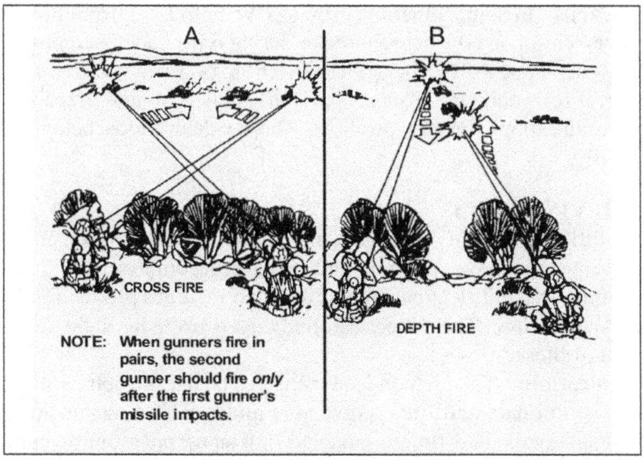


**Figure 8-23. Fire patterns.**

f. **Fire Commands.** The leader can use fire commands to control the gunner's rate of fire, time of fire, and point of fire. However, when distance dictates radio delivery of fire commands, the gunner will have to divide his concentration between listening to the radio and firing the Dragon. His assistant, if he has one, can listen to the radio and relay the radio commands (Table 8-2).

| ELEMENT | DEFINITION | EXAMPLE |
|---|---|---|
| ALERT | Warns the gunner(s) of a fire mission. | "Enemy in sector, prepare to fire." |
| TARGET DESCRIPTION/ LOCATION | Briefly describes the target location relative to a TRP or the gunner | "BMP, 200 meters right of TRP 1." |
| METHOD OF ENGAGEMENT | Provides instructions for engaging a formation — fire cross or depth fire, fire front to rear, fire left to right or right to left, and so forth. | "Team Alpha, fire from rear to center."<br>"Team Bravo, fire from front to center." |
| EXECUTION | Gives the command to fire. | "Team Alpha, fire."<br>"Team Bravo, stand by." |

**Table 8-2. Fire commands and examples.**

g. **Emergency Fire Signals**. How would the leader control fires at a distance without radio? The next quickest option includes using sectors of fire and alternate signals with pyrotechnics. Units must establish SOPs and practice what communications procedures to use when they lose communications.

h. **Range Cards**. In some situations, such as when moving from position to position to cover the movement of an advancing force or during retrograde operations, preparing and using range cards does not work. Then, TRPs offer the best way to control the distribution of fire. However, if time and circumstances permit, gunners should prepare range cards for primary, alternate, and supplementary positions. These valuable tools help the gunner engage targets successfully.

## 8-14. LIMITED VISIBILITY

The gunner has little opportunity to engage targets in darkness, haze, smoke, or fog. The platoon or company must have a simple and quick procedure or SOP that will help him conduct quick limited visibility engagements and that includes provisions for—

a. **Target Acquisition**. The gunner can find targets using his sight, binoculars, NVDs, or a combination of these.

b. **Communications**. The platoon leader ensures he has the communications means to issue target acquisition data and fire commands rapidly. These means must allow him to disseminate SOPs, signals, and fire commands. In a static position, he can "hot loop" all essential personnel on wire communications.

c. **Illumination**. When natural and battlefield light sources, such as burning vehicles, provide too little light to illuminate the target, the gunner uses the AN/TAS-5 nightsight. Leaders may have responsive artificial sources of illumination they can dedicate to supporting antiarmor fires. Two examples of illumination available to platoon leaders for marking TRPs at night include artillery and mortar illumination. (Chapter 1 discusses AN/TAS-5 operational capabilities.)

APPENDIX A

# TRAIN-THE-TRAINER TASKS FOR THE DGT AND DFTT

*Table A-1 lists train-the-trainer tasks for the Dragon gunnery trainer (DGT) and for the Dragon field tactical trainer (DFTT).*

| TASK | CONDITIONS | STANDARD |
|---|---|---|
| **Prepare the DGT for Operation** | Given a DGT and a 110-volt AC power supply. | Assemble the DGT. |
| **Program the DGT for Operation** | Given a DGT, the correct program disks, and a 110-volt AC power supply. | Program the DGT for the correct firing table. |
| **Operate the DGT** | Given a DGT, the correct program disks, and a 110-volt AC power supply. | Operate the DGT using the correct firing table. |
| **Evaluate Gunner Performance** | Given a DGT, the correct program disks, and a 110-volt AC power supply. | Operate the DGT using the correct firing table. Provide the gunner with an after-action review of his tracking and score. |
| **Conduct Shutdown Procedures for DGT** | Given a DGT, the correct program disks, and a 110-volt AC power supply. | Shut down the DGT. |
| **Maintain the DGT** | Given a DGT, a 110-volt AC power supply, and the proper cleaning materials. | Perform maintenance on the DGT. |
| **Prepare the DFTT for Operation** | Given a DFTT and either a 24-volt DC or a 120- or 240-volt AC power supply. | Assemble the DFTT. |
| **Operate the DFTT** | Given a DFTT and either a 24-volt DC or a 120- or 240-volt AC power supply. | Operate the DFTT using the correct firing table. |
| **Evaluate Gunner Performance** | Given a DFTT and either a 24-volt DC or a 120- or 240-volt AC power supply. | Operate the DFTT using the correct firing table. Provide gunner with an after-action review of his tracking and score. |
| **Conduct Shutdown Procedures for DFTT** | Given a DFTT and either a 24-volt DC or a 120- or 240-volt AC power supply. | Shut down the DFTT. |
| **Maintain the DFTT** | Given a DFTT and either a 24-volt DC or a 120- or 240-volt AC power supply. | Perform maintenance on the DFTT. |

**Table A-1. Train-the-trainer tasks for the DGT and the DFTT.**

APPENDIX B

# DRAGON GUNNERY TRAINER

*This appendix provides technical data for the Dragon gunnery trainer (DGT) and discusses its components and limitations. The DGT and related accessories are shipped and stored in two shipping containers, one for the instructor station and one for the student station (Figure B-1).*

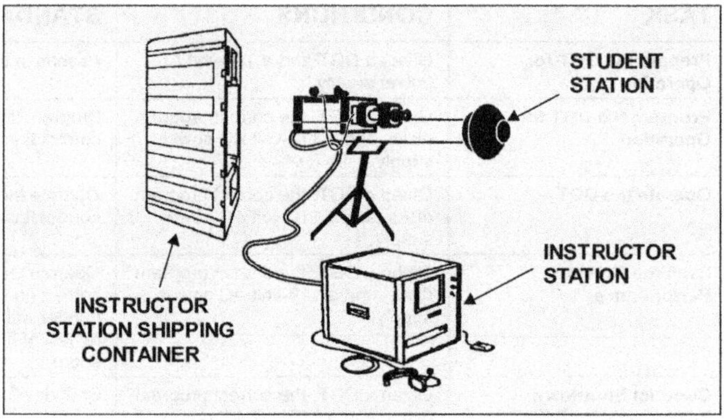


**Figure B-1. Dragon gunnery trainer.**

## Section I. INSTRUCTOR STATION

The instructor station consists of a console, a headset with microphone, a mouse, two power cables, and a shipping container (Figure B-2).

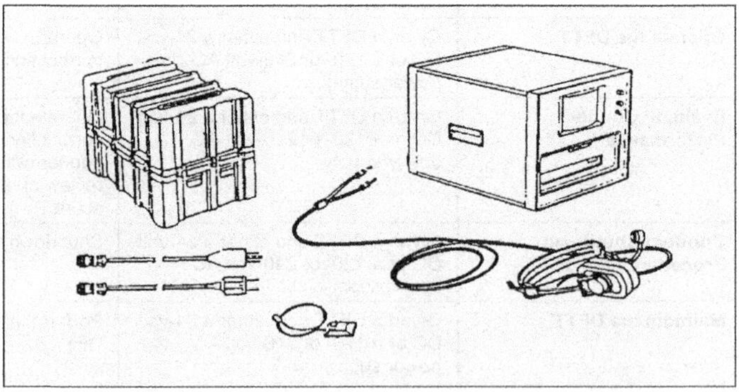

**Figure B-2. Components of instructor station.**

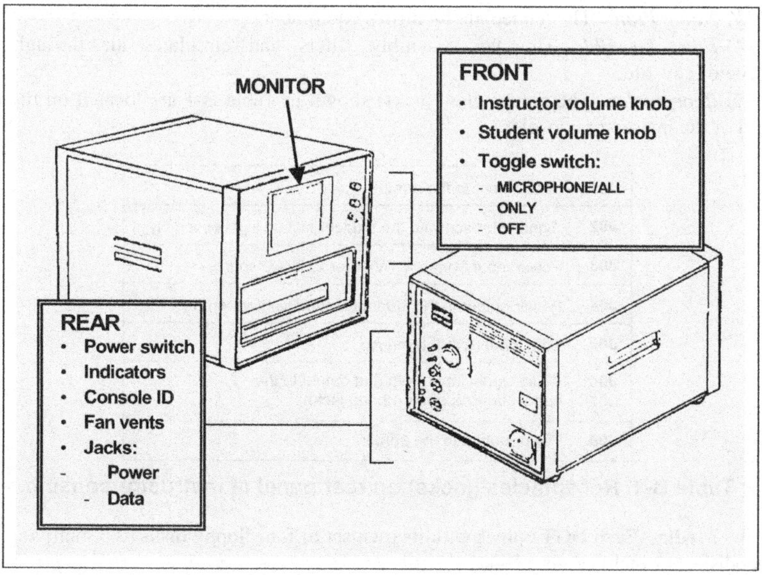


**Figure B-3. Instructor console.**

## B-1. INSTRUCTOR CONSOLE

The trainer programs the DGT instructor console (Figure B-3), which has two carrying handles.

a. **Monitor**. The instructor console displays the mission scenario on its own monitor as well as on the student viewer. This allows the trainer to watch the student's actions during mission execution. Afterwards, the trainer can play back and evaluate the student's performance.

b. **Front Panel**. From the front panel of the instructor console, the trainer controls the volume of his own and the student's headsets (INSTRUCTOR VOLUME and STUDENT VOLUME). The toggle switch on the front panel has three settings:

(1) ***MICROPHONE/ALL***. This setting allows both trainer and student to hear the DGT mission.

(2) ***ONLY***. This setting allows only the trainer to hear the mission.

(3) ***OFF***. This setting is self-explanatory.

c. **Rear Panel**. The rear panel of the instructor console contains the power switch, indicators, console identification data, fan vents, and power and electronic data transmission receptacles (jacks).

(1) ***ON/OFF Switch and Indicator***. The ON/OFF switch controls power to the instructor console; when the switch is set to ON, the indicator light comes on.

(2) ***Hour Meter Indicator***. The hour meter shows how long the trainer has been in use.

(3) ***Name Plate***. The nameplate identifies equipment.

(4) ***Fan Assembly***. The fan assembly filters and circulates air through the instructor console.

(5) ***Receptacles***. The receptacles (jacks) shown in Table B-1 are located on the rear panel of the instructor console:

| | |
|---|---|
| **J01** | Data output to the printer. |
| **J02** | Signal interface with the student station sight assembly. |
| **J03** | Power input from a 110VAC or 220VAC source. |
| **J04** | Power output to the student station sight assembly. |
| **J05** | Data input from the mouse. |
| **J06** **J07** | Audio signal input from and output to the headset/microphone (phone jack). |
| **J08** | Power output to the printer. |

**Table B-1. Receptacles (jacks) on rear panel of instructor console.**

d. **Media**. Each DGT comes with its own set of four floppy disks (3.5 inch) and two videodisks (12 inch) (Figure B-4).

(1) ***Floppy Disks***. These are standard 3.5-inch floppy disks.

(a) *Program Disk*. This disk contains the DGT simulation software.

(b) *Diagnostic Disk*. This disk contains the self-test for the DGT.

(c) *Mission Data Disks*. Each of these two numbered floppy disks corresponds to one of the similarly marked 12-inch videodisks.

(2) ***Videodisks***. Two 12-inch videodisks store the DGT's audiovisuals. Each is numbered and color-labeled like one of the two mission data disks. Each videodisk comes in a protective jacket.

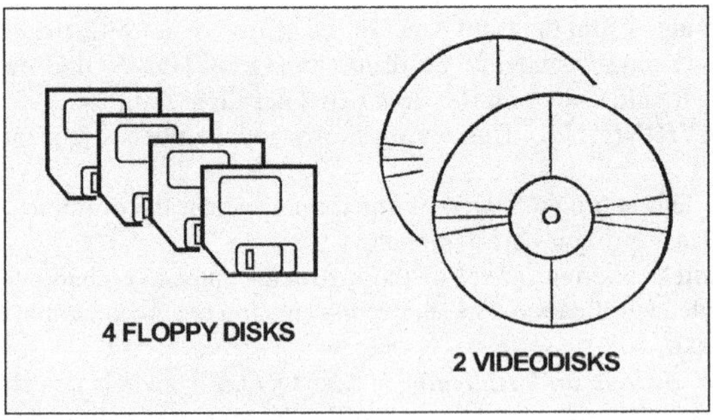


**Figure B-4. DGT media.**

**B-2. HEADSET-MICROPHONE**

The instructor uses the standard communication headset with boom microphone (Figure B-5). This text will simply refer to "the headset." The headset allows the trainer to hear the audio portion of the simulation and to feed back instructions to the student. The headset's 6-foot long cable has two phone plugs that plug into jacks on the instructor console.

**B-3. MOUSE**

The mouse (Figure B-5) is a hand-controlled device the trainer uses to move the cursor on the video monitor. Using the mouse, the trainer controls the cursor, selects options, and controls exercises. To select an option, the instructor positions the cursor on it and clicks the left mouse button once (the right button does not work with this program). The connector on the mouse's 4-foot long cable plugs into a jack on the instructor console.

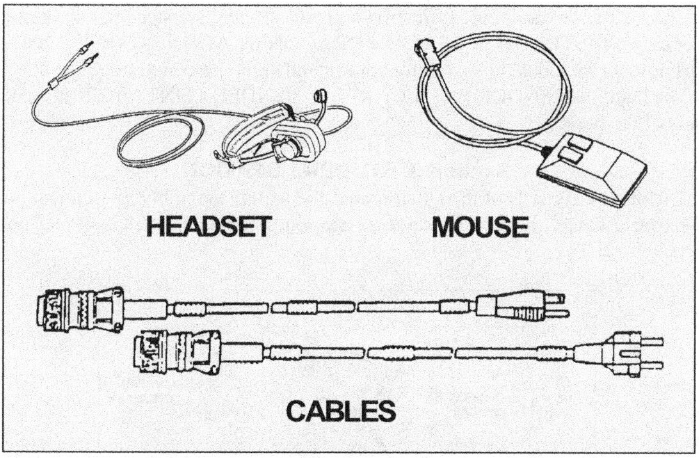


**Figure B-5. Headset, mouse, and cables.**

**B-4. POWER CABLES**

The DGT comes with two power cables (Figure B-5): one for use with 110VAC power (cable W2) and the other for use with 220VAC power (cable W1). Each is 8 feet long.

**B-5. INSTRUCTOR STATION SHIPPING CONTAINER**

Instructor station shipping containers (Figure B-6, page B-5) are made from olive drab-colored polyethylene. Soldiers can stack these containers four deep. Each has specially shaped inserts to prevent damage to the components. The shipping container resists dust and water. The base and cover each have two carrying handles, one on each end. The container has four latches on each side and three latches along each end.

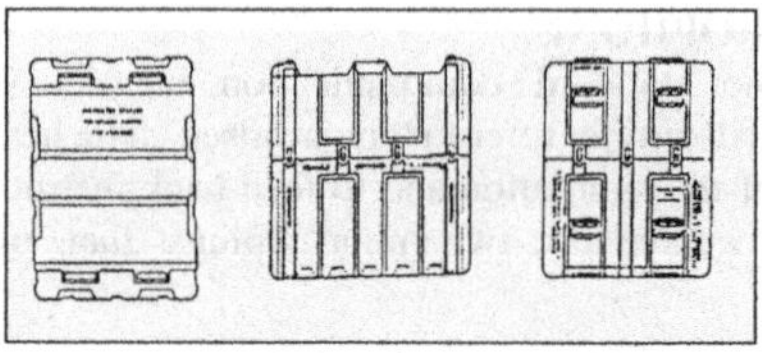

**Figure B-6. Instructor console shipping container.**

a. The container bears the stenciled message on each end in two places: CAUTION, 198 POUNDS, 4-MAN LIFT (MALE ONLY).

b. The shipping container has an identification number and a part number stenciled on the right side of the base along with an arrow icon: REUSABLE CONTAINER P/N 13363801.

c. The instructor station identification and part number are stenciled on the top of the container cover: INSTRUCTOR STATION DRAGON/DRAGON INDOOR P/N 13363800. This part number includes the instructor station and shipping container.

d. The caution HANDLE WITH CARE, REUSABLE CONTAINER is stenciled on both ends of the base.

## Section II. STUDENT STATION

The student station consists of the Dragon unit, the weight assembly, the student headset, a special-purpose cable, the student station's shipping container, disks, and an operator's manual (Figure B-7).

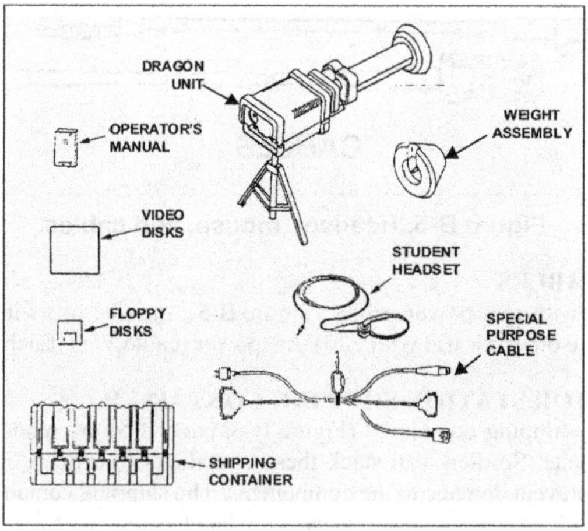


**Figure B-7. Student station components.**

## B-6. DRAGON UNIT

The Dragon unit (Figure B-8) consists of a sight assembly, bipod assembly and retaining strap, and launcher assembly. The student views the mission through the sight assembly. The retaining strap secures the bipod assembly during storage. The bipod assembly supports the launcher and sight assemblies. The launcher assembly simulates the Dragon launch tube.

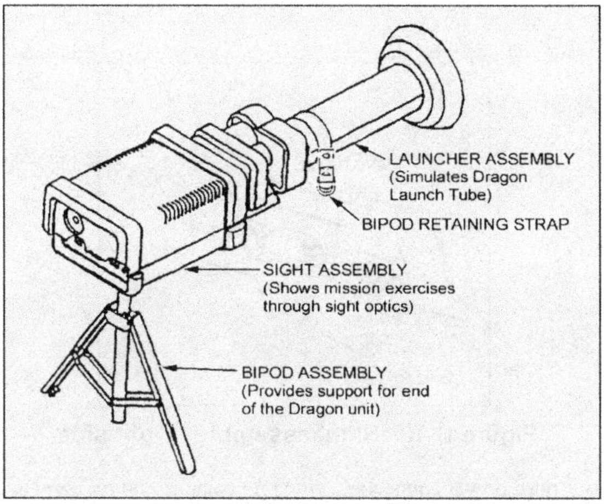


**Figure B-8. Dragon unit.**

a. **Sight Assembly**. The DGT sight assembly (Figure B-9) operates as either the daysight or nightsight, depending on the mission loaded at the instructor console. The sight displays the mission visual signals. The sight assembly attaches to the Dragon launch assembly.

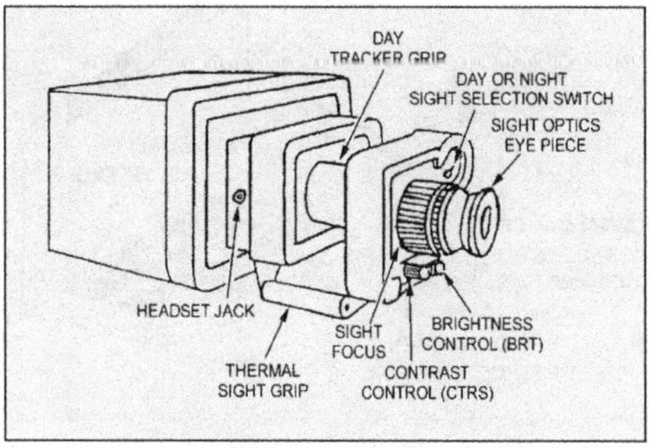


**Figure B-9. Sight assembly.**

(1) The sight assembly has a daysight grip, a switch to select the daysight or nightsight, an eyepiece, a brightness control (BRT), a contrast control (CTRS), a sight focus ring, a thermal sight grip, and a headset jack. The eyepiece simulates both the daysight and nightsight eyepieces.

(2) On its right side (Figure B-10), the sight assembly has a safety and a right-hand trigger grip. Before the student can pull the trigger, he must first depress and hold the safety with his thumb.

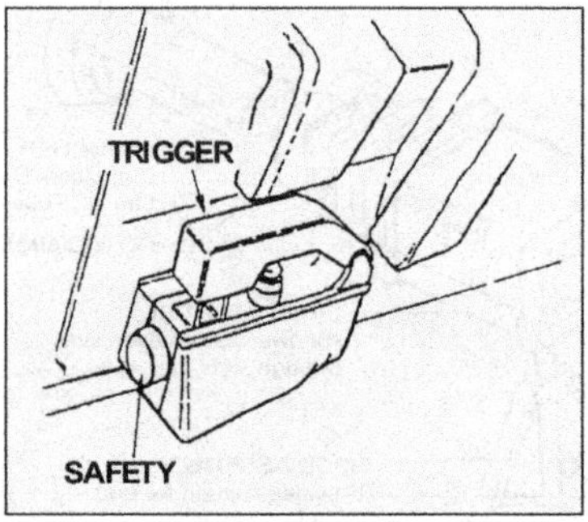


**Figure B-10. Sight assembly—right side.**

(3) The sensor port, power connectors, video receptacle, and recessed analog, digital, and audio jacks are located on the rear panel of the sight assembly (Figure B-11). The trainer may record launch tube movements from the reflector assembly via a sensor port. A CAUTION label designates the sight assembly as a sensitive electronic device.

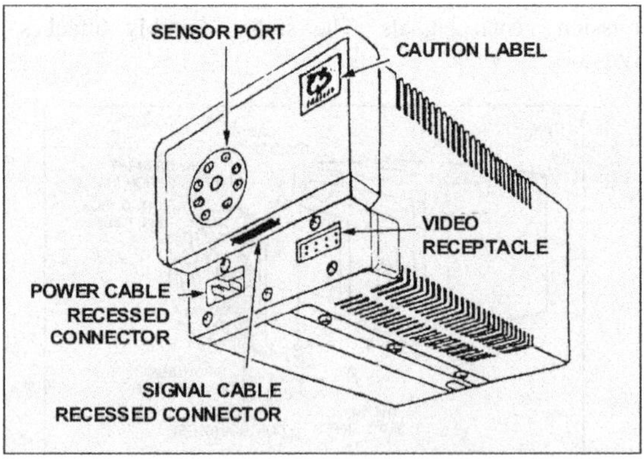


**Figure B-11. Sight assembly—rear panel.**

b. **Bipod Assembly**. The bipod assembly (Figure B-12), which supports the launcher and sight assemblies, attaches to the front end of the launcher assembly.

c. **Launcher Assembly**. This assembly houses the pull-down sensor (Figure B-12). The sensor measures the force with which the student is pulling down on the launch tube. The weight-drop assembly and bipod attach to the launcher assembly.

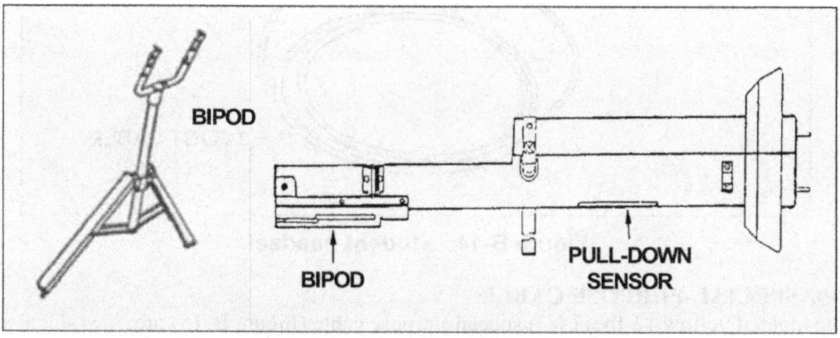


**Figure B-12. Bipod and launcher.**

## B-7. WEIGHT ASSEMBLY

The weight assembly (Figure B-13) drops to simulate the effect that occurs when the Dragon missile leaves the launch tube. A pin in the center of the weight assembly attaches it to the rear of the launch tube. The weight assembly drops 0.6 seconds after the soldier depresses the trigger.

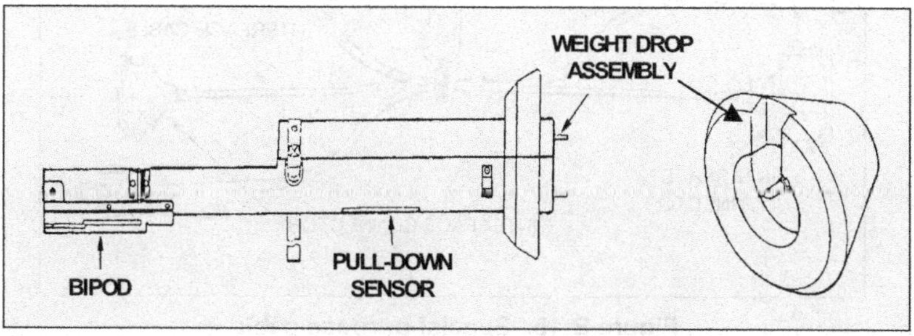


**Figure B-13. Launcher and weight assembly.**

## B-8. HEADSET

The headset (Figure B-14, page B-5) allows the student to hear the instructor's commands as well as the audio portion of the mission. A 3-foot long cable and a headphone plug connect the headset to the sight assembly.

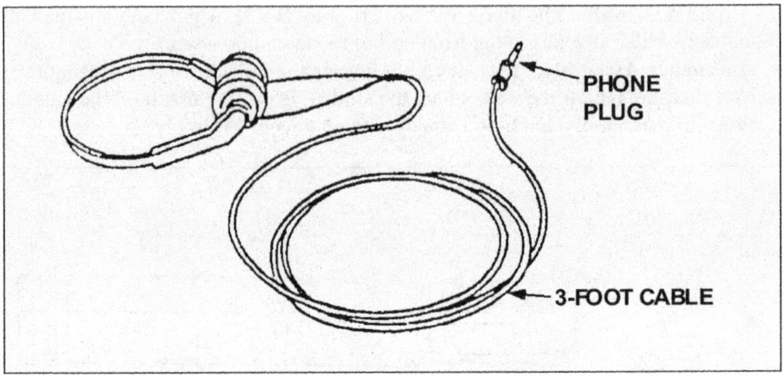


**Figure B-14. Student headset.**

## B-9. SPECIAL-PURPOSE CABLE

The molded yellow 12-foot (W3) special-purpose cable (Figure B-15) provides signal and power connections between the instructor console and the sight assembly.

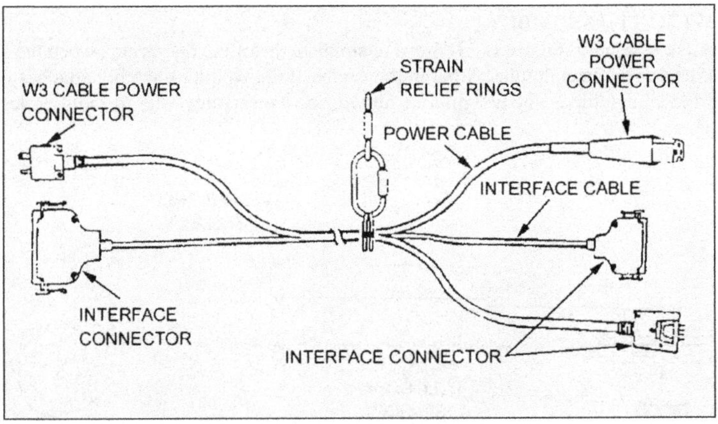


**Figure B-15. Special-purpose cable.**

## B-10. STUDENT STATION SHIPPING CONTAINER

The olive drab-colored, polyethylene, student station shipping containers (Figure B-16, page B-10) have specially shaped inserts to protect the components from damage. The shipping containers resist dust and water, and they stack four deep. The container base has two carrying handles and three latches on each end and seven latches along each side (TM 9-1425-484-10).

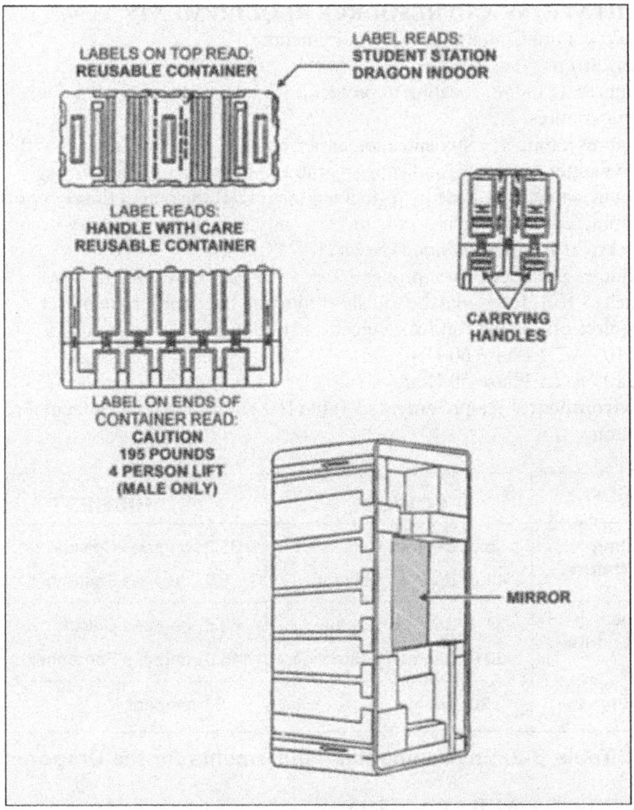


**Figure B-16. Student station shipping container.**

a. **Ends**. The lifting requirement, CAUTION: 195 POUNDS 4 PERSON LIFT (MALE ONLY), is stenciled on each end of the container.

b. **Cover**. The cover has the following labels:

(1) The part number for entire student station (including the shipping container): STUDENT STATION DRAGON INDOOR P/N 13364400.

(2) Part number for shipping container cover only: REUSABLE CONTAINER P/N 13364390.

(3) The warning HANDLE WITH CARE; REUSABLE CONTAINER.

(4) An arrow icon that shows which side goes upward.

c. **Inside Cover**. A mirror is located inside the cover of the student station shipping container.

## .B-11. LIMITATIONS AND RESOURCE REQUIREMENTS

The DGT has certain limitations and requirements.

a. **Limitations**. The DGT—

(1) Requires an indoor location to protect it from humidity, blowing dust and sand, and extreme temperatures.

(2) Requires a trainer, a student, another person. The other person adjusts the bipod legs, replaces the weight assembly, and supports the DGT between firings.

(3) Requires an X-feet wide by Y-feet tall mirror. (If the one in the cover of the student station shipping container breaks or is missing, any mirror this size may be substituted.)

b. **Electrical Requirements**. The DGT—

(1) Requires grounded (three-pronged) extension cords, because its power cables, which are 96 inches (8 feet) long, may be too short to reach the appropriate outlet.

(2) Requires one of the following electrical outlets:

- 110VAC 1 Phase 60 Hertz.
- 220VAC 1 Phase 50 Hertz.

c. **Environmental Requirements**. Table B-2 shows the environmental requirements for the Dragon.

| | LOWEST | HIGHEST |
|---|---|---|
| **Operating Temperature** | 5.0 degrees Celsius | 35.0 degrees Celsius |
| | 41.0 degrees Fahrenheit | 98.0 degrees Fahrenheit |
| **Storage Temperature** | -34.4 degrees Celsius | 62.8 degrees Celsius |
| | -30.0 degrees Fahrenheit | 145.0 degrees Fahrenheit |
| **Humidity** | 20 percent | 80 percent |

**Table B-2. Environmental requirements for the Dragon.**

## B-12. OPERATION AND FUNCTION

The trainer programs the target. Using the DGT mouse and controls, he selects a target and its missions. He also chooses from several options that affect the target's level of difficulty. The target will appear in both the trainer's and the firer's viewers. The firer's viewer has crosshairs. He moves the DGT launcher, placing the crosshairs on the target. He keeps the crosshairs on the target until the missile explodes. Through the headset, he will hear the simulated sounds of the Dragon: the firing of thrusters (when he adjusts his aim) and the blast itself. He will see the impact and know at once whether he hit or missed the target. Throughout this training, he can hear the trainer's commands as well as the battlefield sounds of small arms and artillery fire. At the end of each mission, the trainer's screen displays the firer's score. He can replay or save the mission for future use. If the trainer sets up a printer to use with the DGT, he can print a copy of the mission performance results.

> **DANGER**
>
> **EVEN "LOW VOLTAGES" PRESENT A HAZARD. IN SOME CONDITIONS, VOLTAGE AS LOW AS 50 VOLTS COULD KILL YOU.**

### B-13. SAFETY

Wherever the DGT is used, commanders and leaders must develop local directives and SOPs to supplement existing safety precautions. For the protection of soldiers, these documents should spell out locations; safety requirements; individual responsibilities and soldier's limits; and sequence of operations. Everyone must follow the safety precautions. The DGT uses high voltage. Failure to observe safety precautions could cause death. Before using this equipment, you must know the following rules:

a. Never try to operate electronic equipment alone. Be sure someone else is nearby who knows the operation and hazards of the equipment and can give you first aid.

b. Never touch the metal pins when connecting or disconnecting the power cords.

c. On each piece of equipment, learn which areas carry high voltages. Also, be careful not to contact high-voltage connections when installing or operating this equipment.

d. Before working inside the equipment, turn off the power. Before touching anything, ground anything that might carry a high voltage.

APPENDIX C

# DRAGON FIELD TACTICAL TRAINER

*Leaders use the outdoor DFTT system to train advanced field tracking. This appendix provides technical information about the DFTT and discusses its components and limitations.*

## C-1. RETROREFLECTOR

The DFTT tracks a small retroreflector mounted on the target vehicle. This retroreflector can be placed on virtually any vehicle or item. Only the number of retroreflectors available restricts the number of targets.

## C-2. FEEDBACK

During the simulated flight of the missile, the DFTT provides feedback to the firer. As the missile travels downrange, the DFTT displays the missile's alignment with the target. At the end of the mission, the DFTT displays whether the missile hit or missed; if a hit, it shows where the missile struck the target.

## C-3. MULTIPLE INTEGRATED LASER ENGAGEMENT SYSTEM

Although trainers can use MILES with the DFTT, doing so is impractical. If trainers use this method of training, they will need a retroreflector and MILES belt. The MILES laser requires a clean LOS to the retroreflector. This means that trainers must not hide the retroreflector behind any obstruction such as a tree or brush pile. Leaders should not count MILES training as force-on-force training.

## C-4. SCENARIO

The DFTT also allows the trainer to adjust the difficulty of the scenario to the firer's proficiency by changing—

- The target kill zone.
- The amount of target obscuration caused by the rocket blast.
- The weight lost after the gunner fires a round.

## C-5. CAPABILITIES AND LIMITATIONS

The DFTT can be used either during daylight hours or periods of limited visibility.

a. **Capabilities.**

(1) DFTT provides feedback to the firer during the simulated flight of the missile. It shows how the missile aligns with the target as it travels downrange.

(2) After each mission, the DFTT displays whether or not the missile hit the target and, if so, where on the target it hit.

b. **Limitations.**

(1) Only the number of retroreflectors available to the unit restricts the number of targets presented to the firer.

(2) The afternoon or evening before scheduled training, trainers must recharge the DFTT computer from a 110-volt outlet for at least five hours.

## C-6. SAFETY

Local commanders and leaders should work together to write local directives and SOPs to supplement safety precautions. The SOPs should state individual responsibilities, safety requirements, distance limits for soldiers and explosives, training locations, sequence of operations, equipment needed for handling munitions, and protection of soldiers. Directives or SOP must clearly designate each person's responsibilities in any operation involving explosives. The unit must observe safety precautions.

## C-7. FIRING TABLES

Basic instructional Firing Tables 5 and 6 help new gunners learn about the DFTT.

**WARNING**

**This equipment uses laser light. Always follow laser range safety procedures in AR 385-63 and TB MED 524. Remember—*under the following conditions,* looking directly at a laser light can blind you permanently:**

- **For long periods of time.**
- **At close range.**
- **Through binoculars, telescope, or periscope at a range of less than 154 meters.**

**DANGER**

**THIS EQUIPMENT USES HIGH VOLTAGES AND HIGH-EXPLOSIVE BLAST SIMULATORS.**

**YOU MUST FOLLOW ALL SAFETY PRECAUTIONS WHEN USING THIS EQUIPMENT. FAILURE TO DO SO COULD CAUSE YOU OR SOMEONE ELSE TO LOSE YOUR HEARING, SUFFER TRAUMATIC INJURY, OR DIE.**

**FOR INFORMATION ABOUT ARTIFICIAL RESPIRATION, SEE FM 21-11.**

**1. *HIGH VOLTAGES.***

- **FAILURE TO FOLLOW SAFETY PRECAUTIONS COULD RESULT IN DEATH ON CONTACT.**
- ***DO NOT* TOUCH METAL PINS WHEN CONNECTING OR DISCONNECTING POWER CORDS OR SYSTEM CABLES.**
- ***DO NOT* TRY TO DISASSEMBLE THE COMPUTER.**
- ***DO NOT* USE THE EQUIPMENT IF THE CATHODE RAY TUBE (CRT) OR CABLE IS DAMAGED.**

**2. *HIGH-EXPLOSIVE BLAST SIMULATORS.***

- **WITHIN 70 METERS OF A LOADED BLAST SIMULATOR, WEAR HEARING PROTECTION.**
- **HANDLE BLAST SIMULATORS CAREFULLY. IF POSSIBLE, WEAR FULL HELMET AND GLOVES.**
- **KEEP BLAST SIMULATORS AWAY FROM FIRE.**
- **NEVER HANDLE A DAMAGED BLAST SIMULATOR.**
- **STAY CLEAR OF THE DANGER ZONE AT ALL TIMES.**

APPENDIX D

# FIRING TABLES FOR DRAGON PRECISION GUNNERY TRAINING DEVICES

*The Dragon PGTS replaces the LES/LET trainers. The PGTS includes the Dragon gunnery trainer (DGT) and the Dragon field tactical trainer (DFTT). Up to ten gunners can fire simultaneously at any vehicle or stationary target equipped with a mounted retroreflector. The system provides extensive individual feedback after each engagement.*

## D-1. DGT BASIC INSTRUCTIONAL FIRING TABLES

The basic instructional firing tables for the DGT include Firing Tables 1 through 4.

a. **Firing Table 1.** This firing table introduces the new gunner to the DGT. The gunner will engage stationary targets using the daysight. The hit zone will be set at 100 percent. Students will be allowed to rest for three to five minutes between events. The student will learn—

(1) To use the proper force when pulling down on the Dragon.

(2) To control his breathing.

(3) To aim the weapon properly under different obscuration settings from the sitting position.

| | |
|---|---|
| TASK | Engage stationary and moving targets. |
| CONDITIONS | Given a DGT, DGT disk number 14, LCE, protective mask, and helmet. |
| STANDARD | Breathe and aim properly while in the sitting position. |

b. **Firing Table 2.** The students must be able to track the moving targets with a smooth transition between targets. Trainers will set the hit zone to 100 percent. Students will be allowed to rest for three to five minutes between events.

| | |
|---|---|
| TASK | Engage stationary and moving targets. |
| CONDITIONS | Given a DGT, DGT disk number 14, LCE, protective mask, and helmet. |
| STANDARD | Apply proper breath control and aiming procedures while firing from the sitting position. |

c. **Firing Table 3.** Trainers will set the hit zone to 100 percent. Students will be allowed to rest for three to five minutes between events.

| | |
|---|---|
| TASK | Engage moving targets. |
| CONDITIONS | Given a DGT, DGT disk number 14, LCE, protective mask, and helmet. |
| STANDARD | Apply proper breath control and aiming procedures while in the sitting position. |

d. **Firing Table 4.** The table introduces the new gunner to using the DGT with multiple targets and to using both the daysight and nightsight with different obscuration settings. Students will be required to track while wearing a protective mask. Trainers will set the hit zone to 100 percent. Students will be allowed to rest for three to five minutes between events.

| | |
|---|---|
| TASK | Engage multiple targets. |
| CONDITIONS | Given a DGT, DGT disk number 14, LCE, protective mask, and helmet. |
| STANDARD | Apply proper breath control and aiming procedures while in the sitting position. |

e. **Scorecard**. Use DA Form 7445-R, Dragon Gunnery Trainer (DGT) Scorecard, Moving and Stationary Targets, to score Firing Tables 1 through 4.

(1) This blank, reproducible form may be copied from the back of this manual onto 8 1/2 by 11-inch paper. It is also available on the Army Electronic Library (AEL) CD-ROM (EM0001) and at the USAPA website located at (http://www.usapa.army.mil).

(2) Write the soldier's personal data, the unit and point number, and the date as indicated. Complete the remainder of the form as follows. If you need help, see the example completed scorecard in Figure D-1, page D-5:

- ***ENG.*** Engagement numbers are preprinted on the scorecard.
- ***MSN.*** Mission numbers are preprinted on the scorecard.
- ***OBSC.*** Obscuration level is preprinted on scorecard.
- ***HIT.*** Scorer marks this block if student hits the target.
- ***MISS.*** Scorer marks this block if student misses the target.
- ***TRACKING SCORE.*** Scorer marks the student's tracking score.
- ***THRUSTERS.*** Scorer marks how many thrusters fired.

## D-2. DFTT BASIC INSTRUCTIONAL FIRING TABLES

The basic instructional firing tables for the DFTT include Firing Tables 5 and 6.

a. **Firing Table 5.** The gunner will engage multiple targets using both the daysight and nightsight under varying obscuration settings. Students will be allowed to rest for three to five minutes between events.

| | |
|---|---|
| TASK | Engage multiple targets. |
| CONDITIONS | Given a DFTT, flak vest, LCE, protective mask, helmet, and hearing protection. |
| STANDARD | Fire for familiarization only. |

b. **Firing Table 6.** Students will be allowed to rest for three to five minutes between events.

| | |
|---|---|
| TASK | Engage stationary, moving, and multiple targets. |

| | |
|---|---|
| CONDITIONS | Given a DFTT, flak vest, LCE, protective mask, helmet, and hearing protection. |
| STANDARD | Fire for familiarization only. |

c. **Scorecard**. Use DA Form 7446-R, Dragon Field Tactical Trainer (DFTT) Scorecard, Multiple Targets, to score Firing Tables 5 and 6.

(1) This blank, reproducible form may be copied from the back of this manual onto 8 1/2 by 11-inch paper. It is also available on the Army Electronic Library (AEL) CD-ROM (EM0001) and at the USAPA website located at (http://www.usapa.army.mil).

(2) Write the soldier's personal data, the unit and point number, and the date as indicated. Complete the remainder of the form as follows. If you need help, see the example completed scorecard in Figure D-2, page D-6:

- ***TGT***. This box is already completed.
- ***RANGE***. For each target fired, enter range to target.
- ***OBS***. For each target fired, enter level of obscuration.
- ***HIT KILL***. Check this box if student hit and killed the target.
- ***HIT NO KILL***. Check this box if student hit but did not kill the target.
- ***MISS***. Check this box if student missed the target.
- ***MISSILE GROUNDED***. Check this box if the missile grounded.
- ***EXCEEDED RANGE***. Check this box if the missile exceeded range.
- ***THRUSTERS***. Write the number of thrusters that fired.
- ***SCORER'S SIGNATURE AND DATE.*** Sign and date the form.

## D-3. DGT QUALIFICATION FIRING TABLES

The qualification firing tables for the DGT include Firing Tables 7 and 8.

a. **Firing Table 7.** Soldier practices firing the DGT. This table is for practice qualification only. It evaluates the gunner's ability to engage stationary and moving targets using both the daysight and nightsight. Some targets are more difficult to hit than others.

| | |
|---|---|
| TASK | Engage stationary, moving, and multiple targets. |
| CONDITIONS | Given a DGT, DGT disk number 14, LCE, a protective mask, and helmet. |
| STANDARD | Kill 16 out of 20 target vehicles. |

b. **Firing Table 8.** This table is for qualification. It evaluates the gunner's ability to engage stationary and moving targets using both the daysight and nightsight. Some targets are more difficult to hit than others.

| | |
|---|---|
| TASK | Engage stationary, moving, and multiple targets to qualify. |
| CONDITIONS | Given a DGT, DGT disk number 15, LCE, protective mask, and helmet. |
| STANDARD | Kill 16 out of 20 target vehicles. |

c. **Scorecard**. Use DA Form 7447-R, Dragon Gunnery Trainer (DGT) Scorecard, Multiple Targets, to score Firing Tables 7 and 8.

(1) This blank, reproducible form may be copied from the back of this manual onto 8 1/2 by 11-inch paper. It is also available on the Army Electronic Library (AEL) CD-ROM (EM0001) and at the USAPA website located at (http://www.usapa.army.mil).

(2) Write the soldier's personal data, the unit and point number, and the date as indicated. Complete the remainder of the form as follows. If you need help, see the example completed scorecard in Figure D-3, page D-7:

(a) *ENG*. Engagement number is preprinted on the scorecard.

(b) *MSN*. Mission number is preprinted on the scorecard.

(c) *OBSC*. Level of obscuration is preprinted on the scorecard.

(d) *TEST*. The soldier must score at least 16 in order to qualify on the Dragon. He gets three chances to do so. If the soldier misses 5 engagements, he will not be able to hit the required sixteen targets. He must move on to the first retest, which another scorer oversees.

— *GO*. Check in this column if student scores a hit.

— *NO GO*. Check in this column if student misses.

(e) *RETEST (1)*. The same conditions apply to this test. If the soldier again fails to hit sixteen targets, he moves on to his last retest.

— *GO*. Check in this column if student scores a hit on the retested target.

— *NO GO*. Check in this column if student misses the retested target.

(f) *RETEST (2)*. A third scorer oversees this test. If the soldier misses five (fails to hit sixteen) on this last test, he fails to qualify on the trainer.

— *GO*. Check in this column if student scores a hit on the retested target.

— *NO GO*. Check in this column if student misses the retested target.

(g) *TS*. Enter tracking score.

(g) *TH*. Enter number of thrusters fired.

d. **Scorecard.** On the scoring scale shown to the left of the legend, mark the box next to the scoring category the soldier achieved.

# DRAGON GUNNERY TRAINER (DGT) SCORECARD

## MOVING AND STATIONARY TARGETS

For use of this form see FM 3-23.24 (FM 23-24); the proponent agency is TRADOC

DATA REQUIRED BY PRIVACY ACT OF 1974. AUTHORITY: 10 USC 3012(g)/Executive Order 9397. PRINCIPLE PURPOSE(S): Aids in individual training on targets at various ranges. ROUTINE USE(S): To evaluate individual proficiency. SSN used for positive identification purposes only. MANDATORY OR VOLUNTARY DISCLOSURE AND EFFECT ON INDIVIDUAL NOT PROVIDING INFORMATION: Voluntary. Individuals not providing information cannot be rated or scored on a mass basis.

| LAST NAME | FIRST NAME | MI | SSN | UNIT NO. | POINT NO. | DATE (YYYYMMDD) |
|---|---|---|---|---|---|---|
| Frost | Martin | W | 431-19-4824 | 2-29 IN | 1 | 20010801 |

**FIRING TABLE 1**
STATIONARY AND MOVING TARGETS
DISK: DGT #14
PERCENT HIT ZONE SETTING 50 75 100 %

EVENT A. OPTICAL SIGHT, WEIGHT LOSS OR PULL-DOWN: AIMING POINT AND TARGET ASPECT.

| ENG | MSN | OBSC | HIT | MISS | TRACKING SCORE | THRUSTERS |
|---|---|---|---|---|---|---|
| 1 | 1 | 0.0 | X | | 66 | 11 |
| 2 | 2 | 0.0 | | X | 9 | 14 |
| 3 | 3 | 0.5 | X | | 89 | 15 |
| 4 | 3 | 1.0 | X | | 90 | 23 |
| 5 | 4 | 2.0 | X | | 100 | 16 |

EVENT B. OPTICAL SIGHT: INTRODUCTION TO OBSCURATION AND MOVING.

| ENG | MSN | OBSC | HIT | MISS | TRACKING SCORE | THRUSTERS |
|---|---|---|---|---|---|---|
| 1 | 8 | 0.5 | X | | 63 | 23 |
| 2 | 9 | 1.0 | X | | 91 | 28 |
| 3 | 10 | 0.5 | | X | 25 | 16 |
| 4 | 11 | 0.5 | X | | 65 | 9 |
| 5 | 12 | 0.5 | | X | [illegible] | 28 |

**FIRING TABLE 2**
STATIONARY AND MOVING TARGETS
DISK: DGT #14
PERCENT HIT ZONE SETTING 50 75 100 %

EVENT A. OPTICAL SIGHT: STATIONARY AND MOVING TARGETS.

| ENG | MSN | OBSC | HIT | MISS | TRACKING SCORE | THRUSTERS |
|---|---|---|---|---|---|---|
| 1 | 8 | 0.5 | X | | [illegible] | 23 |
| 2 | 5 | 0.5 | X | | [illegible] | 28 |
| 3 | 18 | 0.0 | X | | 36 | 28 |
| 4 | 19 | 0.0 | X | | 49 | 16 |
| 5 | 20 | 0.0 | X | | 77 | 12 |

EVENT B. BRIGHTNESS AND CONTRAST.

| ENG | MSN | OBSC | HIT | MISS | TRACKING SCORE | THRUSTERS |
|---|---|---|---|---|---|---|
| 1 | 26 | 0.0 | X | | 1 | 7 |
| 2 | 27 | 0.0 | | X | 34 | 15 |
| 3 | 28 | 0.0 | | X | 0 | 22 |
| 4 | 29 | 0.0 | X | | 10 | 20 |
| 5 | 30 | 0.0 | X | | 0 | 15 |

**FIRING TABLE 3**
MOVING TARGETS
DISK: DGT #14
PERCENT HIT ZONE SETTING 50 75 100 %

EVENT A. OPTICAL SIGHT.

| ENG | MSN | OBSC | HIT | MISS | TRACKING SCORE | THRUSTERS |
|---|---|---|---|---|---|---|
| 1 | 8 | 1.0 | X | | 50 | 27 |
| 2 | 9 | 1.0 | X | | 22 | 28 |
| 3 | 10 | 1.0 | X | | 0 | 14 |
| 4 | 11 | 1.0 | X | | 41 | 9 |
| 5 | 12 | 1.0 | | X | 54 | 30 |

EVENT B. THERMAL SIGHT.

| ENG | MSN | OBSC | HIT | MISS | TRACKING SCORE | THRUSTERS |
|---|---|---|---|---|---|---|
| 1 | [illegible] | 0.0 | X | | 25 | 7 |
| 2 | 27 | 0.0 | | X | 0 | 14 |
| 3 | 28 | 0.0 | X | | 0 | 20 |
| 4 | 29 | 0.0 | X | | 0 | 14 |
| 5 | 30 | 0.0 | X | | 45 | 16 |

**FIRING TABLE 4**
INTERMEDIATE TARGETS
DISK: DGT #14
PERCENT HIT ZONE SETTING 50 75 100 %

EVENT A. OPTICAL SIGHT.

| ENG | MSN | OBSC | HIT | MISS | TRACKING SCORE | THRUSTERS |
|---|---|---|---|---|---|---|
| 1 | 7 | 0.5 | X | | 88 | 19 |
| 2 | 9 | 1.0 | X | | 0 | 27 |
| 3 | 10 | 1.5 | X | | 0 | 15 |
| 4 | 11 | 0.5 | X | | 78 | 9 |
| 5 | 12 | 1.0 | X | | 62 | 30 |

EVENT B. OPTICAL AND THERMAL SIGHT.

| ENG | MSN | OBSC | HIT | MISS | TRACKING SCORE | THRUSTERS |
|---|---|---|---|---|---|---|
| 1 | 13 | THERMAL | X | | 20 | 18 |
| 2 | 26 | THERMAL | X | | 56 | 7 |
| 3 | 28 | THERMAL | | X | 0 | 22 |
| 4 | 29 | THERMAL | X | | 56 | 17 |
| 5 | 30 | THERMAL | X | | 6 | 15 |

EVENT C. OPTICAL SIGHTS, MOPP, PROTECTIVE MASK FAMILIARIZATION.

| ENG | MSN | OBSC | HIT | MISS | TRACKING SCORE | THRUSTERS |
|---|---|---|---|---|---|---|
| 1 | 7 | 0.0 | X | | 84 | 19 |
| 2 | 9 | 0.0 | X | | 0 | 28 |
| 3 | 10 | 0.0 | X | | 76 | 15 |
| 4 | 11 | 0.0 | X | | 60 | 9 |
| 5 | 12 | 0.0 | X | | 61 | 29 |

LEGEND: ENG - ENGAGEMENT
MSN - MISSION
OBSC - OBSCURATION

REMARKS:

SCORER'S NAME (PRINT) SGT Jon Flores

SCORER'S SIGNATURE SSG Jon Flores

DA FORM 7445-R, DEC 2001

EXAMPLE

**Figure D-1. Example completed DA Form 7445-R, Dragon Gunnery Trainer (DGT) Scorecard (Moving and Stationary Targets), Firing Tables 1 through 4.**

# DRAGON FIELD TACTICAL TRAINER (DFTT) SCORECARD
## MULTIPLE TARGETS

**TARGET SIZE SETTING: 1 2 3**

For use of this form see FM 3-23.24 (FM 23-24); the proponent agency is TRADOC.

DATA REQUIRED BY PRIVACY ACT OF 1974. AUTHORITY: 10 USC 3012(g)/Executive Order 9397. PRINCIPLE PURPOSE(S): Aids in individual training on targets at various ranges. ROUTINE USE(S): To evaluate individual proficiency. SSN used for positive identification purposes only. MANDATORY OR VOLUNTARY DISCLOSURE AND EFFECT ON INDIVIDUAL NOT PROVIDING INFORMATION: Voluntary. Individuals not providing information cannot be rated or scored on a mass basis.

| LAST NAME | FIRST NAME | MI | SSN | UNIT NO | POINT NO. | DATE (YYYYMMDD) |
|---|---|---|---|---|---|---|
| Frost | Martin | W | 431-19-4824 | 2-29 IN | 1 | 20011206 |

EXAMPLE

### FIRING TABLE 5

MULTIPLE TARGETS, FAMILIARIZATION ONLY

EVENT A. OPTICAL SIGHT, SITTING.

| TGT | RANGE | *OBS | HIT KILL | HIT NO KILL | MISS | MISSILE GROUNDED | EXCEEDED RANGE | THRUSTERS |
|---|---|---|---|---|---|---|---|---|
| 1 | 450 | 2 | | X | | | | 12 |
| 2 | 507 | 2 | | | X | | | 5 |
| 3 | 852 | 2 | X | | | | | 16 |
| 4 | 757 | 2 | X | | | | | 8 |
| 5 | 510 | 2 | | X | | | | 19 |

EVENT B. OPTICAL SIGHT, STANDING.

| TGT | RANGE | *OBS | HIT KILL | HIT NO KILL | MISS | MISSILE GROUNDED | EXCEEDED RANGE | THRUSTERS |
|---|---|---|---|---|---|---|---|---|
| 1 | 912 | 2 | | | | | X | 30 |
| 2 | 805 | 2 | | X | | | | 26 |
| 3 | 650 | 2 | | | X | | | 18 |
| 4 | 450 | 2 | X | | | | | 7 |
| 5 | 375 | 2 | X | | | | | 6 |

EVENT C. THERMAL SIGHT, SITTING.

| TGT | RANGE | *OBS | HIT KILL | HIT NO KILL | MISS | MISSILE GROUNDED | EXCEEDED RANGE | THRUSTERS |
|---|---|---|---|---|---|---|---|---|
| 1 | 450 | 0 | X | | | | | 11 |
| 2 | 523 | 0 | | X | | | | 14 |
| 3 | 648 | 0 | X | | | | | 17 |
| 4 | 719 | 0 | | | X | | | 20 |
| 5 | 863 | 0 | | X | | | | 21 |

EVENT D. THERMAL SIGHT, STANDING.

| TGT | RANGE | *OBS | HIT KILL | HIT NO KILL | MISS | MISSILE GROUNDED | EXCEEDED RANGE | THRUSTERS |
|---|---|---|---|---|---|---|---|---|
| 1 | 920 | 0 | | | | | X | 30 |
| 2 | 412 | 0 | | | | X | | 14 |
| 3 | 557 | 0 | | X | | | | 17 |
| 4 | 650 | 0 | | X | | | | 20 |
| 5 | 787 | 0 | X | | | | | 25 |

### FIRING TABLE 6

STATIONARY, MOVING, AND MULTIPLE TARGETS, FAMILIARIZATION ONLY

EVENT A. OPTICAL SIGHT, SITTING.

| TGT | RANGE | *OBS | HIT KILL | HIT NO KILL | MISS | MISSILE GROUNDED | EXCEEDED RANGE | THRUSTERS |
|---|---|---|---|---|---|---|---|---|
| 1 | 445 | 2 | X | | | | | 10 |
| 2 | 574 | 2 | | | X | | | 15 |
| 3 | 635 | 2 | | X | | | | 18 |
| 4 | 747 | 2 | | | | | X | 30 |
| 5 | 856 | 2 | | | | X | | 24 |

EVENT B. OPTICAL SIGHT, STANDING.

| TGT | RANGE | *OBS | HIT KILL | HIT NO KILL | MISS | MISSILE GROUNDED | EXCEEDED RANGE | THRUSTERS |
|---|---|---|---|---|---|---|---|---|
| 1 | 560 | 2 | | X | | | | 17 |
| 2 | 701 | 2 | X | | | | | 8 |
| 3 | 737 | 2 | X | | | | | 21 |
| 4 | 869 | 2 | X | | | | | 24 |
| 5 | 910 | 2 | | X | | | | 29 |

EVENT C. THERMAL SIGHT, SITTING.

| TGT | RANGE | *OBS | HIT KILL | HIT NO KILL | MISS | MISSILE GROUNDED | EXCEEDED RANGE | THRUSTERS |
|---|---|---|---|---|---|---|---|---|
| 1 | 854 | 0 | X | | | | | 23 |
| 2 | 570 | 0 | X | | | | | 12 |
| 3 | 451 | 0 | X | | | | | 9 |
| 4 | 912 | 0 | X | | | | | 28 |
| 5 | 787 | 0 | X | | | | | 21 |

EVENT D. THERMAL SIGHT, STANDING.

| TGT | RANGE | *OBS | HIT KILL | HIT NO KILL | MISS | MISSILE GROUNDED | EXCEEDED RANGE | THRUSTERS |
|---|---|---|---|---|---|---|---|---|
| 1 | 633 | 0 | X | | | | | 19 |
| 2 | 557 | 0 | X | | | | | 17 |
| 3 | 480 | 0 | | X | | | | 14 |
| 4 | 857 | 0 | X | | | | | 23 |
| 5 | 920 | 0 | | | | | X | 30 |

REMARKS:

SCORER'S NAME (PRINT) SSG Jonathan Flores

SCORER'S SIGNATURE SSG Jonathan Flores

* OBSCURATION

DA FORM 7446-R, DEC 2001 USAPA V1.00

**Figure D-2. Example completed DA Form 7446-R, Dragon Field Tactical Trainer (DFTT) Scorecard (Multiple Targets), Firing Tables 5 and 6.**

# DRAGON GUNNERY TRAINER (DGT) SCORECARD

## MULTIPLE TARGETS

For use of this form see FM 3-23.24 (FM 23-24); the proponent agency is TRADOC.

DATA REQUIRED BY PRIVACY ACT OF 1974. AUTHORITY: 10 USC 3012(g)/Executive Order 9397. PRINCIPLE PURPOSE(S): Aids in individual training on targets at various ranges. ROUTINE USE(S): To evaluate individual proficiency. SSN used for positive identification purposes only. MANDATORY OR VOLUNTARY DISCLOSURE AND EFFECT ON INDIVIDUAL NOT PROVIDING INFORMATION: Voluntary. Individuals not providing information cannot be rated or scored on a mass basis.

| LAST NAME | FIRST NAME | MI | SSN | UNIT NO. | POINT NO. | DATE (YYYYMMDD) |
|---|---|---|---|---|---|---|
| Frost | Martin | W | 431-19-4824 | 2-29 IN | I | 20011206 |

**FIRING TABLE 7**

DGT PRACTICE QUALIFICATION
DISK: DGT #14
PERCENT HIT ZONE SETTING
50 % 75 % 100 %

EVENT A. OPTICAL SIGHT.

| ENG | MSN | OBSC | TEST GO | TEST NO GO | TEST TS | TEST TH |
|---|---|---|---|---|---|---|
| 1 | 1 | 0.5 | X | | 97 | 11 |
| 2 | 2 | 1.5 | X | | 0 | 11 |
| 3 | 7 | 0.5 | X | | 86 | 18 |
| 4 | 8 | 1.5 | X | | 76 | 23 |
| 5 | 9 | 0.5 | X | | 40 | 28 |
| 6 | 10 | 1.5 | X | | 0 | 15 |
| 7 | 11 | 0.5 | X | | 57 | 9 |
| 8 | 12 | 1.5 | X | | 35 | 29 |
| 9 | 13 | 0.5 | | X | 0 | 17 |
| 10 | 14 | 1.5 | X | | [illegible] | 14 |

EVENT B. THERMAL SIGHT.

| ENG | MSN | OBSC | TEST GO | TEST NO GO | TEST TS | TEST TH |
|---|---|---|---|---|---|---|
| 1 | 26 | 0.0 | X | | 49 | [illegible] |
| 2 | 27 | 0.0 | X | | 0 | 10 |
| 3 | 28 | 0.0 | | X | [illegible] | [illegible] |
| 4 | 29 | 0.0 | X | | 40 | 21 |
| 5 | 30 | 0.0 | X | | 19 | 16 |
| 6 | 26 | 0.0 | X | | 0 | 8 |
| 7 | 27 | 0.0 | | X | 0 | 12 |
| 8 | 28 | 0.0 | X | | 23 | 22 |
| 9 | 29 | 0.0 | X | | 0 | 10 |
| 10 | 30 | 0.0 | X | | 22 | 15 |

**FIRING TABLE 8**

DGT QUALIFICATION
DISK: DGT #15
PERCENT HIT ZONE SETTING
50 % 75 % 100 %

EVENT A. OPTICAL SIGHT.

| ENG | MSN | OBSC | TEST GO | TEST NO GO | TEST TS | TEST TH | RETEST GO | RETEST NO GO | RETEST TS | RETEST TH | RETEST GO | RETEST NO GO | RETEST TS | RETEST TH |
|---|---|---|---|---|---|---|---|---|---|---|---|---|---|---|
| 1 | 1 | 0.5 | X | | [illegible] | 9 | | | | | | | | |
| 2 | 6 | 1.5 | X | | 77 | 20 | | | | | | | | |
| 3 | 7 | [illegible] | [illegible] | | [illegible] | [illegible] | | | | | | | | |
| 4 | 4 | [illegible] | X | | 83 | 19 | | | | | | | | |
| 5 | [illegible] | 0.5 | X | | 78 | [illegible] | | | | | | | | |
| 6 | 11 | [illegible] | | X | 0 | [illegible] | | | | | | | | |
| 7 | 12 | 0.5 | X | | 82 | 23 | | | | | | | | |
| 8 | [illegible] | [illegible] | X | | [illegible] | 24 | | | | | | | | |
| [illegible] | 14 | [illegible] | X | | 0 | 26 | | | | | | | | |
| [illegible] | [illegible] | [illegible] | X | | 65 | 23 | | | | | | | | |

EVENT B. THERMAL SIGHT.

| ENG | MSN | OBSC | TEST GO | TEST NO GO | TEST TS | TEST TH | RETEST GO | RETEST NO GO | RETEST TS | RETEST TH | RETEST GO | RETEST NO GO | RETEST TS | RETEST TH |
|---|---|---|---|---|---|---|---|---|---|---|---|---|---|---|
| 1 | 17 | 0.0 | X | | 63 | 17 | | | | | | | | |
| 2 | 18 | 0.0 | X | | 48 | 22 | | | | | | | | |
| 3 | 19 | 0.0 | X | | 64 | 24 | | | | | | | | |
| 4 | 20 | 0.0 | X | | 24 | 30 | | | | | | | | |
| 5 | 21 | 0.0 | X | | 71 | 13 | | | | | | | | |
| 6 | 22 | 0.0 | X | | 55 | 20 | | | | | | | | |
| 7 | 23 | 0.0 | X | | 2 | 11 | | | | | | | | |
| 8 | 24 | 0.0 | X | | 56 | 22 | | | | | | | | |
| 9 | 25 | 0.0 | X | | 65 | 16 | | | | | | | | |
| 10 | 26 | 0.0 | X | | 69 | 13 | | | | | | | | |

EXAMPLE

LEGEND: ENG - ENGAGEMENT; MSN - MISSION; OBSC - OBSCURATION; TH - THRUSTERS (NARROW COLUMN); TS - TRACKING SCORE (NARROW COLUMN)

REMARKS:

**SCORING (CHECK ONE):**

- EXPERT [X] 19 to 20
- SHARPSHOOTER [ ] 17 to 18
- MARKSMAN [ ] 16
- UNQUALIFIED [ ] Less than 16

SCORER'S NAME (PRINT): SSG Jonathan Flores

SCORER'S SIGNATURE: SSG Jonathan Flores

DA FORM 7447-R, DEC 2001 USAPA V1.00

**Figure D-3. Example completed DA Form 7447-R, Dragon Gunnery Trainer (DGT) Scorecard (Multiple Targets), Firing Tables 7 and 8.**

APPENDIX E

# M113 ARMORED PERSONNEL CARRIER

*This appendix is included for all the units that still have M113 APCs.*

## E-1. FIRING FROM THE M113 APC

You can fire the Dragon from the M113 in three ways.

a. The best way to fire the Dragon from a stationary M113 APC in a hull-defilade position is to use the M175 mount (Figure E-1), though field-expedient methods can be used.

**Figure E-1. Firing the Dragon from an M113 APC.**

(1) Adjust the commander's chair in the APC so that, when you assume the firing position, both feet are flat on the commander's chair and your knees and body are straight.

(2) Keep your feet apart at a comfortable angle.

(3) Ensure that your left arm and hand exert maximum rearward pressure on the elevation damper. This action pulls the cradle down against your shoulder (Figure E-2).

**Figure E-2. M175 firing position.**

(4) Position your right hand on the M175 mount firing mechanism the same as you would on the sight's firing mechanism.

- Right thumb on safety.
- Index, middle, and third fingers on the trigger lever.
- Little finger forward of the trigger mechanism.

**WARNING**

**To prevent a missile fired from an M175-mounted Dragon from dipping and flying into the ground, press heavily back and down on the elevation damper.**

b. If not using the M175 mount, the gunner stands in the track commander's hatch, places the Dragon's bipod legs in the grill on top of the APC, and fires the weapon (Figure E-3). He could also sit on the closed cargo hatch and fire the Dragon as if he were on the ground (Figure E-4, page E-3).

c. The gunner can also fire the Dragon from a dismounted position, using either the M3 or M122 machine gun tripod (Figure E-5, page E-3).

(1) ***M3 Tripod***. Sit along the left side of the tripod to ensure that no portion of your body is forward of the muzzle. Sight, aim, and fire the Dragon the same as when it is mounted on an APC.

(2) ***M122 Tripod***. With the M122 tripod, put the right leg over the top of the two rear tripod legs. Put the left leg over the top of the front tripod leg. Sight, aim, and fire the same as when mounted on an APC.

**NOTE:** Body position is the most important factor in successful firing and tracking. Providing soldiers with continual feedback during the early phase of practical training will help them form good habits.

**WARNING**

**TO ENSURE THE SAFETY OF THE PERSONNEL IN THE APC, CLOSE THE DRIVER'S AND CARGO HATCHES BEFORE FIRING.**

**Figure E-3. Firing the Dragon from the commander's hatch.**

Figure E-4. Firing from a closed hatch.

Figure E-5. M3 position (left) and M122 tripod position (right).

**E-2. SUPPORT EQUIPMENT (M113-EQUIPPED UNITS ONLY)**

Units equipped with the M113 APC are authorized the following Dragon support equipment: M175 guided missile launcher mounts, M3 or M122 machine gun tripods, and vehicle storage kits.

a. **M175 Mounting Assembly**. The guided missile launcher mount (Figure E-6, page E-4) provides a stable platform for firing the Dragon missile from the M113 APC. The unit track mechanic installs the M175 mounting assembly on the APC. Before turning in the APC for overhaul or repair, the unit track mechanic must remove all support equipment.

(1) The M175 mount is an improved, stable firing and tracking base for the gunner that increases the effectiveness of the weapon system.

(2) The gunner uses the M175 mount to fire the M222 tactical round or the M223 practice round.

(3) The round and sight mate electrically through two connectors on the M175 mount. The remote firing mechanism is located on the right rear of the cradle.

**Figure E-6. M175 guided missile launcher mount.**

(4) Azimuth and elevation dampers on the M175 mount reduce vibrations and help the gunner track more smoothly.

b. **M3/M122 Machine Gun Tripods**. The gunner can use either the M3 or the M122 machine gun tripod (Figure E-7) to fire the Dragon from a ground support position.

c. **Vehicle Storage Kit**. Each M113 APC assigned to a mechanized infantry squad has one of these kits installed (Figure E-8, page E-5). It consists of the following nightsight support equipment:

(1) A nightsight storage rack, located just below the daysight storage case.

(2) A vehicle power conditioner (VPC) to "step down" the power of the 24-volt battery to the required level.

(3) A battery charger rack and a PP-7382/TAS battery charger also operate from the APC power source.

(4) Three storage racks for coolant cartridge containers and one storage rack for a battery container are located under the personnel seats.

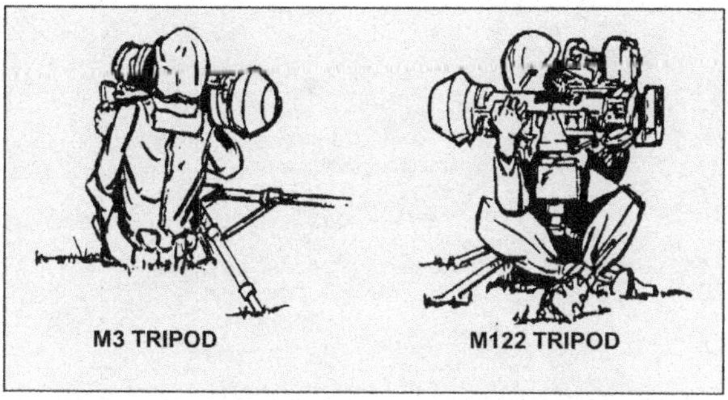


**Figure E-7. M3/M122 machine gun tripods.**

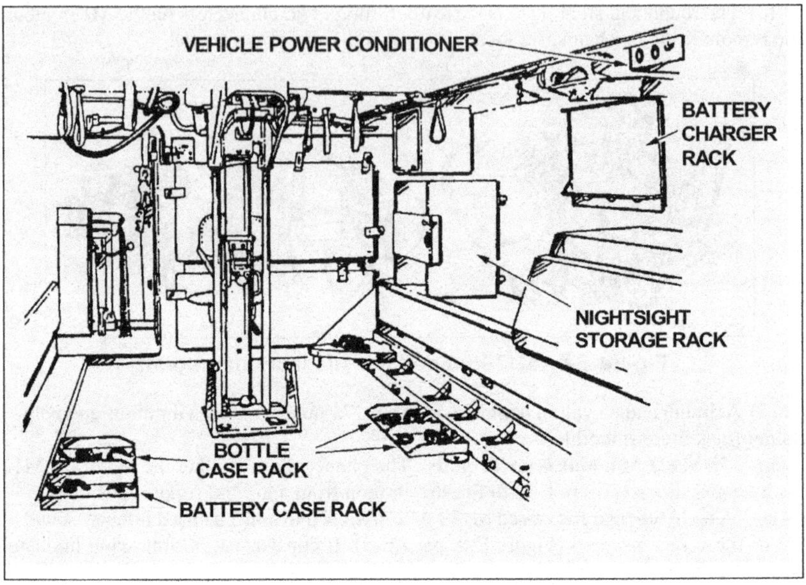


Figure E-8. Vehicle storage kit.

APPENDIX F
# SAFETY

*Commanders and leaders supplement this appendix and the other published materials referenced in it with local directives and SOPs. They can require soldiers to obtain and use this appendix and the published materials referenced in it, or they can incorporate this appendix and the materials it references.*

*In addition, local directives and SOPs must supplement this appendix with the following information:*

- *Individual responsibilities.*
- *Additional or more stringent local safety requirements.*
- *Distance limitations for soldiers and explosives.*
- *Locations and sequence of operations.*
- *Equipment needed for handling munitions.*
- *Soldier protection.*

*Finally, in any operation involving explosives, directives and SOPs should clearly designate and explain each soldier's responsibilities.*

## F-1. SAFETY PRECAUTIONS

After removing the Dragon's front shock absorber and desiccant package, all soldiers must strictly observe *additional* safety precautions to prevent personnel from tampering with the encased missile.

a. Before they may fire the Dragon, leaders and soldiers must know and follow these guidelines:

(1) Avoid looking at the sun, flares, lasers, or searchlights while looking through the sight. Doing so magnifies their effect and could severely burn your eyes.

(2) The warhead will not arm within 65 meters, so try to avoid firing it within that range.

(3) Due to the risk of an accidental discharge, use extreme care when mating the sight to the round.

(4) Avoid going or placing any equipment or other items forward of the firing position.

(5) Never perform PMCS when the sight is mated to the round.

(6) If you are within 50 meters of a Dragon when it is fired, you must wear your earplugs.

(7) Never fire the Dragon over friendly soldiers or vehicles.

b. The antireflective coating on the AN/TAS-5 infrared lens contains thorium fluoride, which is slightly radioactive. However, this coating material is only hazardous if swallowed or inhaled.

c. The Dragon backblast area extends 50 meters to the rear and 30 meters to the flanks of the launcher (Figure F-1, page F-2). This area is divided into two zones: a danger zone and a caution zone. Trainers must keep both zones clear during training.

(1) The *danger zone* extends 30 meters to the rear of the launcher in a 90-degree cone. Within this area, the blast, flame, and flying debris could cause deaths or serious casualties, so personnel must avoid this zone.

(2) The *caution zone* extends an additional 20 meters to the rear and 30 meters to the flanks of the danger zone. Personnel should try to avoid this area as well. To protect their eyes from flying debris, they *must* face away from the rear of the Dragon. To protect their ears from the blast and overpressure, they must wear earplugs.

d. Due to the danger of the backblast, everyone must take care during all phases of instruction. Team drills, position exercises, and tracking exercises must be conducted with the DGT and DFTT just as though a tactical missile were being fired.

e. After flight, a practice missile can still contain live rocket thrusters. Though the warhead on this type of missile is inert, the live rocket thrusters present a hazard to personnel. The locations of all practice missiles must be reported to EOD so they can dispose of the missiles.

f. During live-fire exercises, safety personnel must stand on either side of the gunner. However, due to the gases, flame, and debris that escape from the front and rear of the Dragon, safety personnel must stand at least 1 meter away from the Dragon.

**WARNING**

**Missile guidance wire, which can hang up on trees or bushes, is thin, strong, and hard to see. Avoid leaving any of it around. Control access to any areas where you think some might remain. Remove any wire that you find, then store or dispose of it properly.**

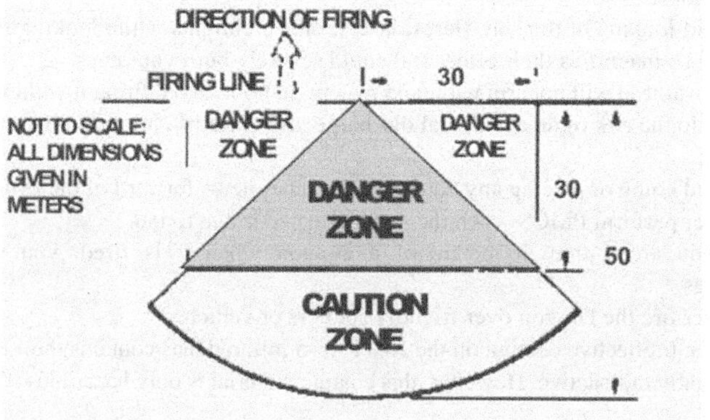


**Figure F-1. Dragon backblast area.**

**WARNINGS**

1. **Wear ear protection devices to prevent ear damage.**
2. **Keep personnel clear during training exercises and nontactical firings. Noise levels and flying debris could cause serious ear or eye damage.**

## F-2. ROUND HANDLING

Improperly or carelessly handling the Dragon round can damage its components and cause the missile to malfunction when it is launched. If the gunner notices any sign that a missile may have been dropped, or if the launcher is deformed or fractured, the gunner should return it to the responsible ammunition personnel for inspection and disposition.

## F-3. FIRING LIMITATIONS

The gunner must not fire the Dragon from within buildings, bunkers, or other enclosures, or within 15 meters (50 feet) of a vertical or nearly vertical backstop. Trainers may be granted waivers from this limitation under the provisions of AR 385-62. In combat, the gunner may have to risk firing the Dragon from an enclosure. If so, he must ensure the enclosure is at least 3 meters by 4.6 meters (10 feet by 15 feet). Also, he must ensure all debris and loose objects are cleared from behind the launch site. When possible, he should ensure all doors and windows are opened. The team should make holes at least 0.6 meter (2 feet) square in the walls and ceilings to allow the backblast and overpressure to escape. Even when the enclosure meets these requirements, the gunner must wear double hearing protection. Also, firing a Dragon from a confined space will almost certainly cause structural damage and create falling debris. It also concentrates the escaping toxic gases and can cause building fires. Any one of these hazards can injure the gunner or team. A Dragon should never be positioned so the gunner has to fire over power lines, through brush or brush fires, or through limbs or other obstructions. Any of these could damage the Dragon's command-link wire and interfere with missile guidance.

## F-4. FIGHTING POSITION

When firing from a fighting position, the gunner should ensure there are no obstructions to his front or rear. These could deflect the launcher's backblast onto him. When firing from a downhill or uphill slope, he must be especially careful to ensure that the angle of the launcher relative to the ground is no more than 20 degrees. When firing from either a hasty or prepared fighting position, the gunner must allow at least 15 centimeters (6 inches) above and below the muzzle of the launcher. This gives the missile fins the room they need to unfold and lock into place.

## F-5. FIRING OVER WATER

When firing the Dragon over salt water, the gunner must avoid firing at targets beyond 300 meters. Salt water can short-circuit the command link wire. Raising the launcher 0.3 meter

(1 foot) increases the distance the Dragon can be fired over water by 100 meters. Fresh water does not affect command-link wire, so the missile normally can be fired over it.

## F-6. TRAINING EQUIPMENT

Safety precautions apply in training just as they do during live firing. For example, during training, soldiers must practice and observe the same backblast precautions they have to observe during live fire (Figure F-1). Even the trainers (DFTT and MILES) generate sound and pressure levels that can damage hearing. Consequently, anyone in or near the firing area must wear proper hearing protection.

APPENDIX G

# GUNNER'S PERFORMANCE TEST

*The trainer sets up the equipment mentioned in the conditions statement, and provides the student with information IAW the evaluation preparation. This appendix lists tasks, conditions, and standards for the gunner's performance test.*

## G-1. EVALUATION PREPARATION

The trainer tells each student that he will be required to perform without error all the actions to complete each task being tested.

## G-2. TEST ADMINISTRATION

The trainer administers the tests and completes a scorecard for each student.

a. Destroy or recycle any copies of the 1990 edition of DA Form 4242-1-R, Dragon Performance Scorecard--they are obsolete. Copy the new, blank, reproducible DA Form 4242-1-R from the back of this manual onto 8 1/2 by 11-inch paper. Soon after the publication of this manual, those with Internet access should be able to download an electronic copy of the revised blank DA Form 4242-1-R from the USAPA web site. Figure G-1 shows an example of the completed form. Instructions for completing it follow:

(1) **LAST NAME**. Write student's last name.

(2) **FIRST NAME**. Write student's first name.

(3) **MI**. Write student's middle initial.

(4) **SSN**. Write student's social security number.

(5) **RANK**. Write student's rank.

(6) **UNIT**. Write student's unit.

(7) **TEST DATE**. For student's first test, mark the GO or NO-GO column for the appropriate training objective.

(8) **RETEST DATES**. For student's second and third tests, mark the GO or NO-GO column for the appropriate training objective.

(9) **REMARKS**. Write your comments about the student's performance.

(10) **QUALIFIED**. Check this column when the student qualifies.

(11) **UNQUALIFIED**. Check this column if the student fails to qualify on the first test or either retest. If the student gets three checks in this column, he fails to qualify on the weapon.

(12) **INSTRUCTOR'S SIGNATURE**. Sign here for each test that you score.

b. Record results on the Dragon scorecard.

c. Provide added training on each task failed, and retest each failed task twice.

d. Eliminate students who fail any task after three tries.

e. Substitute an FHT wherever a task states to use a round.

## G-3. TRAINING OBJECTIVES

Training objectives begin on the page that follows Figure G-1.

a. Training Objective 1, Maintain the M47 Medium Antitank Weapon, page G-3.

b. Training Objective 2, Prepare the M47 for Firing, page G-5.

c. Training Objective 3, Restore the M47 to Carrying Configuration, page G-6.

d. Training Objective 4, Demonstrate Correct Firing Positions, page G-7.

e. Training Objective 5, Determine If Targets Are Engageable, page G-8.

f. Training Objective 6, Perform Immediate-Action Procedures, page G-9.

g. Training Objective 7, Prepare an Antiarmor Range Card, page G-11.

h. Training Objective 8, Identify the Six Major Components of an M47 Fighting Position, page G-12.

I. Training Objective 9, Identify Armored Vehicles, page G-13.

j. Training Objective 10, Explain Emergency Destruction Procedures, page G-14.

k. Training Objective 11, Explain Emergency Decontamination Procedures, page G-15.

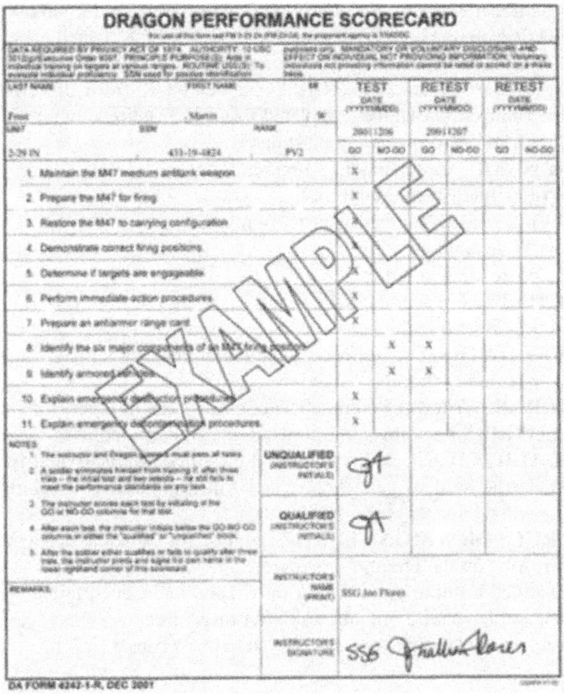
DRAGON PERFORMANCE SCORECARD

| | TEST | | RETEST | | RETEST | |
|---|---|---|---|---|---|---|
| | GO | NO-GO | GO | NO-GO | GO | NO-GO |
| 1. Maintain the M47 medium antitank weapon | X | | | | | |
| 2. Prepare the M47 for firing | X | | | | | |
| 3. Restore the M47 to carrying configuration | X | | | | | |
| 4. Demonstrate correct firing positions | X | | | | | |
| 5. Determine if targets are engageable | X | | | | | |
| 6. Perform immediate action procedures | X | | | | | |
| 7. Prepare an antiarmor range card | X | | | | | |
| 8. Identify the six major components of an M47 firing position | | X | X | | | |
| 9. Identify armored vehicles | | X | X | | | |
| 10. Explain emergency destruction procedures | X | | | | | |
| 11. Explain emergency decontamination procedures | X | | | | | |

EXAMPLE

DA FORM 4242-1-R, DEC 2001

**Figure G-1. Example completed DA Form 4242-1-R, Dragon Performance Scorecard.**

## TRAINING OBJECTIVE 1
## MAINTAIN THE M47 MEDIUM ANTITANK WEAPON

**Conditions**. Given Dragon sights and a round of ammunition.

**Standards**. Conduct a complete preoperational check of the Dragon sights and a round of ammunition.

**Evaluation Preparation**.

*Brief student*: "When I say 'Begin,' you are to conduct a preoperational check of the Dragon sights and a round of ammunition. Do you have any questions?"

*Answer the student's questions*. If the student still does not understand, say, "Do the best you can."

**Instructor's Note**. Because the daysight and nightsight have several parts in common, the student may mention these parts on only one or the other. If he mentions it on either, give him a "GO."

> **WARNING**
> **Never perform a preoperational check on a Dragon sight while you still have it mated to a round of ammunition.**

| Performance Measures | GO | NO-GO | NO-GO | NO-GO |
|---|---|---|---|---|
| 1. **Daysight**.<br>a. Pushed safety plunger.<br>b. Squeezed trigger lever bar.<br>c. Ensured safety was functional and trigger clicked. | | | | |
| 2. **Guide pins**. Looked for the guide pins, ensured there were four of them present, and ensured they were tight. | | | | |
| 3. **Connector**.<br>a. Checked connector cover for rubber seals.<br>b. Ensured connector was clean.<br>c. Ensured connector was undamaged. | | | | |
| 4. **Sight housing**.<br>a. If connector sustained damage, reported this to supervisor.<br>b. Spot-painted as needed. | | | | |
| 5. **Eye guard**. Rotated the eye guard. | | | | |
| 6. **Nightsight**.<br>a. Spot-painted as needed.<br>b. Checked range focus lever.<br>c. Checked for freedom of movement. | | | | |

| Performance Measures | GO | NO-GO | NO-GO | NO-GO |
|---|---|---|---|---|
| 7. **Daysight**. Cleaned the lenses. | | | | |
| 8. **Actuator switch.** | | | | |
| a. Rotated the switch to AIR-BATT check. | | | | |
| b. Checked the coolant gauge (2,500 PSI). | | | | |
| c. Checked the battery monitor on the coolant cartridge. | | | | |
| d. Ensured the reticles were illuminated. | | | | |
| e. Rotated the actuator switch to the ON position. | | | | |
| f. Waited 15 seconds. | | | | |
| g. Removed the lens cover. | | | | |
| 9. **Infrared contrast**. | | | | |
| a. Selected an object with infrared contrast. | | | | |
| b. Adjusted the reticle focus. | | | | |
| c. Adjusted brightness. | | | | |
| d. Adjusted contrast. | | | | |
| e. Adjusted range focus. | | | | |
| f. Rotated the actuator switch to the OFF-LOCK position. | | | | |
| g. Did not turn the actuator switch to the RELEASE position. | | | | |
| h. Replaced the lens cover. | | | | |
| 10. **Ammunition round.** | | | | |
| a. Checked the exterior for dirt, oil, and grease. | | | | |
| b. Checked the exterior for dents, holes, and punctures. | | | | |
| 11. **Raceway**. Ensured raceway was not damaged. | | | | |

**Signature Block**

INSTRUCTOR'S SIGNATURE: ______________________

STUDENT'S INITIALS: GO __________

NO-GO __________

NO-GO __________

NO-GO __________

## TRAINING OBJECTIVE 2
## PREPARE THE M47 FOR FIRING

**Conditions**. Given a Dragon sight in the carrying bag and a round of ammunition in the carrying configuration.

**Standards**. Mate the sight to the round of ammunition IAW the performance measures described herein.

**Evaluation Preparation.**

*Brief student.* "When I say 'Begin,' you are to prepare the Dragon for firing by mating the sight to the round of ammunition. Do you have any questions?"

*Answer the student's questions.* If the student still does not understand, say, "Do the best you can."

| Performance Measures | GO | NO-GO | NO-GO | NO-GO |
|---|---|---|---|---|
| 1. **Prepared the round.**<br>a. Unsnapped bipod-retaining strap.<br>b. Lowered bipod to vertical-lock position. (Shook bipod and ensured it was locked.)<br>c. Depressed friction lock and extended legs about halfway down.<br>d. Removed the electrical connector cover from the round.<br>2. **Prepared the daysight.**<br>a. Removed the sight properly.<br>b. Removed electrical connector cover.<br>c. Checked lenses.<br>3. **Mated the sight to the round.**<br>a. Aligned the guide pins with the sight support assembly and locked the sight in place. (Shook sight to ensure it was locked.)<br>b. Leveled sight picture. (Adjusted the bipod friction lock and the adjusted the foot.) | | | | |

**Signature Block**

INSTRUCTOR'S SIGNATURE: ____________________

STUDENT'S INITIALS: GO __________

NO-GO __________

NO-GO __________

NO-GO __________

TRAINING OBJECTIVE 3

## RESTORE THE M47 TO CARRYING CONFIGURATION

**Conditions**. Given a Dragon sight, carrying bag, and round of ammunition.

**Standards**. Properly return the sight and round to the carrying configuration without damaging the equipment.

**Evaluation Preparation**.

*Brief student.* "When I say 'begin,' you will restore the Dragon to the carrying configuration. Do you have any questions?"

*Answer the student's questions*. If the student still does not understand, say, "Do the best you can."

| Performance Measures | GO | NO-GO | NO-GO | NO-GO |
|---|---|---|---|---|
| 1. **Removed and stored the sight.** | | | | |
| a. Released the spring clip. | | | | |
| b. Removed from support assembly. | | | | |
| c. Removed the sight. | | | | |
| d. Replaced the lens cover. | | | | |
| e. Replaced electrical cover. | | | | |
| f. Placed sight in the carrying bag. | | | | |
| 2. **Restored the round of ammunition.** | | | | |
| a. Replaced the connector cover. | | | | |
| b. Collapsed the bipod legs. | | | | |
| c. Replaced the shock absorber. | | | | |
| d. Secured legs to the round. | | | | |

**Signature Block**

INSTRUCTOR'S SIGNATURE: ____________________

STUDENT'S INITIALS: GO ________

NO-GO ________

NO-GO ________

NO-GO ________

TRAINING OBJECTIVE 4

## DEMONSTRATE CORRECT FIRING POSITIONS

**Conditions**. Given a Dragon sight already mated to a Dragon round.

**Standards**. Correctly assume the three correct Dragon firing positions.

**Evaluation Preparation.**

*Brief student.* "When I say 'Begin,' you will assume the three firing positions for the Dragon. We will test you on all three. After you assume each correct position, tell me which position it is, and I will inform you if you are a GO or NO-GO for that position. Do you have any questions?"

*Answer the student's questions.* If the student still does not understand, say, "Do the best you can."

| Performance Measures | GO | NO-GO | NO-GO | NO-GO |
|---|---|---|---|---|
| 1. **Sitting position.**<br>a. Sat down, feet on bipod. Pushed legs out.<br>b. Leaned forward and placed round on the meaty portion of shoulder, keeping the upper body straight.<br>c. Properly grasped the sight with both hands.<br>d. Rotated head and aligned right eye in the eye guard. Pushed head forward.<br>e. Pulled elbows in and backward, and tried to touch them together on the chest. Removed slack from the bipod. | | | | |
| 2. **Standing supported.**<br>a. Stood with legs spread comfortably apart and straight, but not locked.<br>b. Leaned forward. Rested the abdomen against the edge of the fighting position.<br>c. Placed Dragon on the meaty part of the shoulder. (Placed bipod legs far forward.)<br>d. Placed hands as for the sitting position.<br>e. Rotated head and right eye as in the sitting position.<br>f. Removed slack from the bipod. Pulled down and back. Tried to touch the elbows and drive them into the chest. If there was slack in the bipod, straightened the back slightly. | | | | |
| 3. **Kneeling position.**<br>a. Kneeled with knees spread comfortably apart. | | | | |

| Performance Measures | GO | NO-GO | NO-GO | NO-GO |
|---|---|---|---|---|
| b. Placed the bipod well out in front and removed slack from the bipod by rocking backward. Tried to sit on heels.<br>c. Rotated head and right eye as in the sitting position.<br>d. Placed hand and elbows the same as in the sitting position. | | | | |

**Signature Block**

INSTRUCTOR'S SIGNATURE: ____________________

STUDENT'S INITIALS: GO ________

NO-GO ________

NO-GO ________

NO-GO ________

## TRAINING OBJECTIVE 5
## DETERMINE IF TARGETS ARE ENGAGEABLE

**Conditions**. Given a Dragon sight mated to a round of ammunition and targets (scale or actual targets).

**Standards**. Correctly determine if three targets are in range and engageable.

**Evaluation Preparation.**

*Brief student.* "To your front are armored vehicles at varying (simulated) ranges. When I say 'Begin,' determine whether the targets are in range. Then you must accurately determine the engageability of all three targets. Do you have any questions?"

*Answer the student's questions.* If the student still does not understand, say, "Do the best you can."

| Performance Measures | GO | NO-GO | NO-GO | NO-GO |
|---|---|---|---|---|
| Determine if the target is—<br>1. In range or out of range, engageable or not.<br>**Target Number 1, ___, ___.**<br>2. In range or out of range, engageable or not.<br>**Target Number 2, ___, ___.**<br>3. In range or out of range, engageable or not.<br>**Target Number 3, ___, ___.** | | | | |

**Signature Block**

| | | |
|---|---|---|
| INSTRUCTOR'S SIGNATURE: | | ____________________ |
| STUDENT'S INITIALS: | GO | ________ |
| | NO-GO | ________ |
| | NO-GO | ________ |
| | NO-GO | ________ |

## TRAINING OBJECTIVE 6
## PERFORM IMMEDIATE-ACTION PROCEDURES

**Conditions**. Given a malfunctioning Dragon sight mated to a Dragon round.

**Standards**. Perform the proper immediate-action procedures for hot and cold sight batteries.

**Evaluation Preparation**.

*Brief student.* "When I say 'Begin,' perform immediate-action procedures for the Dragon. Do you have any questions?"

*Answer the student's questions.* If the student still does not understand, say, "Do the best you can."

| Performance Measures | GO | NO-GO | NO-GO | NO-GO |
|---|---|---|---|---|
| 1. **Hot sight battery**. | | | | |
| a. Tried and failed to fire. | | | | |
| b. Immediately resqueezed the trigger for 15 seconds. | | | | |
| c. Announced "Misfire" if round still did not fire. | | | | |
| d. Turned, cautiously felt near the sight battery, and found it to be hot. | | | | |
| e. Removed the sight and placed the round a safe distance away. | | | | |
| f. Obtained a new round and mated the sight to it. | | | | |
| g. Acquired the target, said, "I am continuing my mission" or squeezed the trigger. | | | | |
| h. Squeezed the trigger. | | | | |
| 2. **Cold sight battery**. | | | | |
| a. Tried and failed to fire. | | | | |
| b. Immediately resqueezed the trigger for 15 seconds. | | | | |
| c. Announced "Misfire" if round still did not fire. | | | | |

| Performance Measures | GO | NO-GO | NO-GO | NO-GO |
|---|---|---|---|---|
| c. Turned, cautiously felt near the sight battery, and found it to be cold. | | | | |
| d. Released the sight and moved it forward; reseated the sight on the round. | | | | |
| e. Acquired the target, said, "I am continuing my mission" or squeezed the trigger. | | | | |
| f. When sight again failed to fire, rechecked battery, which was still cold. | | | | |
| g. Removed sight; placed round a safe distance away. | | | | |
| h. Got a new round and said, "I will continue the mission." | | | | |
| i. When round again failed to fire, rechecked battery, which was still cold. | | | | |
| j. Replaced the sight. | | | | |

**Signature Block**

INSTRUCTOR'S SIGNATURE: ____________________

STUDENT'S INITIALS: GO ________

NO-GO ________

NO-GO ________

NO-GO ________

## TRAINING OBJECTIVE 7
## PREPARE AN ANTIARMOR RANGE CARD

**Conditions**. Given the standard range card and squad leader's briefing containing all pertinent data.

**Standards**. Within 15 minutes, complete the antiarmor range card. Include all data from the squad leader's briefing, marginal data, a scaled-down sketch of the terrain, and the data section.

**Evaluation Preparation**.

*Brief student*. "When I say 'Begin,' you will have 15 minutes to complete the antiarmor range card, including the data section and terrain sketch. Do you have any questions?"

*Answer the student's questions*. If the student still does not understand, say, "Do the best you can."

| Performance Measures | GO | NO-GO | NO-GO | NO-GO |
|---|---|---|---|---|
| 1. **Marginal information.**<br>a. Unit designation.<br>b. Magnetic north. | | | | |
| 2. **Terrain sketch.**<br>a. Left limit.<br>b. Right limit.<br>c. Maximum engagement line.<br>d. Terrain features.<br>e. Dead space.<br>f. Target reference point(s).<br>g. Gunner reference point. | | | | |
| 3. **Data section.**<br>a. Position.<br>b. Date.<br>c. Weapon symbol.<br>d. Meter increments.<br>e. Item numbers.<br>f. Direction.<br>g. Range.<br>h. Description.<br>i. Remarks. | | | | |
| 4. **Two copies of the range card.** | | | | |

**Signature Block**

INSTRUCTOR'S SIGNATURE: ______________________

STUDENT'S INITIALS: GO ________

NO-GO ________

NO-GO ________

NO-GO ________

TRAINING OBJECTIVE 8

## IDENTIFY THE SIX MAJOR COMPONENTS OF AN M47 FIGHTING POSITION

**Conditions**. Given a predrawn fighting position.

**Standards**. Within 10 minutes, correctly label the six main components of a fighting position.

**Evaluation Preparation**.

*Brief student.* "When I say 'Begin,' you will have 10 minutes to draw and label the six main components of a Dragon fighting position, IAW the instructions given in the respective performance measure. Do you have any questions?"

*Answer the student's questions.* If the student still does not understand, say, "Do the best you can."

| Performance Measures | GO | NO-GO | NO-GO | NO-GO |
|---|---|---|---|---|
| 1. **Trench.**<br>a. Three M16s long, shaped like an inverted "V."<br>b. Waist deep.<br>c. Waist wide plus 6 inches.<br>2. **Front parapet.**<br>a. One M16 long.<br>b. One M16 wide.<br>c. Two helmets high.<br>3. **Flank parapet.**<br>a. One M16 wide.<br>b. Two helmets high.<br>c. Long enough to provide good flank protection.<br>4. **Bipod trench.**<br>a. Two helmets long.<br>b. One helmet wide.<br>c. Six inches deep (one bayonet blade).<br>5. **Grenade sump.**<br>a. One entrenching tool-length long.<br>b. One entrenching tool-length deep.<br>c. One entrenching tool-blade wide.<br>6. **Overhead cover.**<br>a. Each end of the trench.<br>b. Dug 12 inches deep.<br>c. Eighteen inches longer than the side of the trench.<br>d. Three feet wide.<br>e. The hole under the overhead cover is large enough for one student and extra ammunition. | | | | |

**Signature Block**

INSTRUCTOR'S SIGNATURE: ______________________

STUDENT'S INITIALS: GO ________

NO-GO ________

NO-GO ________

NO-GO ________

## TRAINING OBJECTIVE 9
## IDENTIFY ARMORED VEHICLES

**Conditions.** In a garrison or field environment, given armored vehicles visible in a tactical or simulated tactical setting, a Dragon daysight, and a simulated round of ammunition (FHT), prepared for firing.

**Standards.** Within two minutes, recognize 10 out of 10 vehicles as friendly or enemy.

**Evaluation Preparation.**

*Setup.* The basic combat vehicle identification training aid, GTA 17-2-13, shows multiple views of 30 friendly and enemy armored vehicles. These are the vehicles that every student should know. Show each student the pictures of any six enemy and any four friendly armored vehicles. Show each picture for 10 seconds.

*Brief student.* "I will show you each picture for 10 seconds. During that time, you must state whether the vehicle is friendly or enemy."

*Answer the student's questions.* If the student still does not understand, say, "Do the best you can."

| Performance Measures | GO | NO-GO | NO-GO | NO-GO |
|---|---|---|---|---|
| Recognized 10 out of 10 armored vehicles as friendly or enemy. | | | | |

**Signature Block**

INSTRUCTOR'S SIGNATURE: ______________________

STUDENT'S INITIALS: GO ________

NO-GO ________

NO-GO ________

NO-GO ________

## TRAINING OBJECTIVE 10
# EXPLAIN EMERGENCY DESTRUCTION PROCEDURES

**Conditions**. Given a simulated situation in which capture is imminent, and the order to destroy the sight and rounds of ammunition.

**Standards**. Explain how to destroy Dragon components and ammunition.

**Evaluation Preparation**.

*Brief student*. "When I say 'begin,' you must explain emergency destruction procedures for a Dragon. Do you have any questions?"

*Answer the student's questions*. If the student still does not understand, say, "Do the best you can."

| Performance Measures | GO | NO-GO | NO-GO | NO-GO |
|---|---|---|---|---|
| **NOTE:** Destruction priority: First the sight, then the round. | | | | |
| 1. **Sight.** | | | | |
| a. Destroyed sight by mechanical means. Used an entrenching tool, axe, sledgehammer, or other appropriate item. | | | | |
| b. Destroyed by gunfire. | | | | |
| c. Destroyed by burning. Used flammable material or gasoline. | | | | |
| d. Destroyed by exploding. Used a double-primed explosive charge. | | | | |
| 2. **Round**. Destroyed a round of ammunition by either— | | | | |
| a. *Burning it*. Used flammable material, thermite grenade, or gasoline. | | | | |
| b. *Exploding it*. Used a double-primed explosive charge. (Squad leader will prepare explosives.) | | | | |
| c. *Launching it*. Fired the missile, then drove over and crushed the expelled launcher. | | | | |

**Signature Block**

INSTRUCTOR'S SIGNATURE: ______________________

STUDENT'S INITIALS: GO __________

NO-GO __________

NO-GO __________

NO-GO __________

## TRAINING OBJECTIVE 11
# EXPLAIN EMERGENCY DECONTAMINATION PROCEDURES

**Conditions**. Given the Dragon sights and a round of ammunition (for all tasks, use the FHT instead of an actual round), M8 paper, M256 detector kit, M258A1 kit, hot, soapy water, lens tissue, camel's-hair brush, and ethyl alcohol.

**Standards**. Correctly explain the decontamination procedures for the Dragon sights and round of ammunition.

**Evaluation Preparation**.

*Brief student*. "When I say 'Begin,' you will correctly explain how to decontaminate the Dragon sights and a round of ammunition. Do you have any questions?"

*Answer the student's questions*. If the student still does not understand, say, "Do the best you can."

**Instructor's Note**. Decide whether the Dragon needs immediate decontamination. If the gunner is already wearing MOPP4 and the threat is imminent, decontamination must be postponed. If trainers plan to displace the gunner to a noncontaminated area, you must decide whether to fire the missile immediately or try to decontaminate it for transport. (Inform the student of the type of decontamination: nuclear, chemical, or biological.)

| Performance Measures | GO | NO-GO | NO-GO | NO-GO |
|---|---|---|---|---|
| 1. **Nuclear decontamination**. | | | | |
| a. Thoroughly brushed clothing and equipment to remove fallout (away from the position). | | | | |
| b. Turned over the soil in the immediate area around the position. | | | | |
| c. Verified decontamination with the AN/PDR-27 radiacmeter. | | | | |
| 2. **Biological decontamination**. | | | | |
| a. Washed the Dragon with warm, soapy water. | | | | |
| b. Dried the Dragon. | | | | |
| c. Used normal cleaning procedures for the lenses. | | | | |
| 3. **Chemical decontamination**. | | | | |
| a. Checked with the M256 detector kit. | | | | |
| b. Checked with M8 paper. | | | | |
| c. Removed or blotted excess agent from all components. | | | | |
| d. Left Dragon exposed to air and sun. | | | | |
| e. Used hot, soapy water (if available). Did not submerge the Dragon or sight. | | | | |

| Performance Measures | GO | NO-GO | NO-GO | NO-GO |
|---|---|---|---|---|
| f. Used the M258A1 kit on parts the gunner needed to touch or that touched the gunner. | | | | |
| g. Used normal cleaning procedures for the lenses. | | | | |

**Signature Block**

INSTRUCTOR'S SIGNATURE: ____________________

STUDENT'S INITIALS: GO ________

NO-GO ________

NO-GO ________

NO-GO ________

APPENDIX H

# DECONTAMINATION AND DESTRUCTION PROCEDURES

*Units develop and maintain decontamination procedures so they can accomplish their missions in possible NBC environments. A unit will destroy its Dragons only in the combat zone, only on order or IAW command policy, and only to prevent the enemy from capturing and using them.*

## Section I. AGENTS

Decontamination means removal or neutralization of a hazardous level of NBC contamination from personnel and materials. Decide whether or not to decontaminate a Dragon contaminated by NBC agents. Remove nuclear contamination (fallout) as soon as possible. However, as long as you wear the required protective equipment, you can carry and fire a Dragon contaminated with chemical or biological agents. After doing so, however, swap your MOPP gear out IAW unit SOP. Chemical and biological contaminants soak into the exposed rubber-like and plastic parts of the Dragon. You will not be able to decontaminate these parts completely. Instead of wasting your time on them, decontaminate the other parts of the Dragon that you actually touch.

### H-1. CHEMICAL CONTAMINANTS

Air out components to eliminate chemical agents.

a. Remove or blot extra liquid agents from all components.

b. Place the equipment outdoors in the sun.

c. Use M8 detector paper and an M256 detector kit to check the Dragon periodically, until it is safe to handle (FM 3-5).

d. To quickly reduce chemical contamination on Dragon surfaces, use hot, soapy water (do not submerge) or an M258A1 kit.

e. Use a camel's hair brush, ethyl alcohol, and lens paper to gently clean the lens.

f. Avoid using DS2 and STB, the standard decontaminants, because they can damage the Dragon weapon system's sensitive electronic components.

### H-2. BIOLOGICAL AGENTS

Eliminate biological agents as follows:

a. Wash the Dragon with warm, soapy water.

b. Dry the Dragon.

c. Clean the lens with lens-cleaning solvents (or ethyl alcohol) and lens paper.

### H-3. NUCLEAR AGENTS

Do not remove nuclear agents until all fallout has stopped. Then remove them as follows:

a. Move away from the fighting position and brush clothing and equipment thoroughly to remove fallout.

b. Decontaminate individual equipment by brushing, wiping, and scrubbing.

c. Turn over the soil to decontaminate the immediate area around the position.
d. Use the AN/PDR-27 radiacmeter to verify that the Dragon is decontaminated.

## Section II. DESTRUCTION GUIDANCE

If soldiers cannot evacuate the Dragons, then the unit should destroy first the sights, then the rounds. Destroying the same component in all weapons prevents the enemy from assembling a complete Dragon. Soldiers destroy Dragons by mechanical or explosive means, gunfire, or fire. Some procedures require the use of explosives that may not be authorized. The commander must decide whether to issue these and related materials. If the situation requires the destruction of Dragons, the commander orders the unit to destroy them where the debris will obstruct the enemy but will not endanger friendly soldiers. Each organization and installation that uses, maintains, or stores Dragon missiles or sights should specify in its SOP how to destroy them. The destruction plan should allow for any situation. It should state priorities and methods of destruction; it should provide clear instructions for each method of destruction; and it should specify the quantities of explosives required.

### H-4. DESTRUCTION OF SIGHT

When you destroy the sight, use one of the following methods:

a. **Explosives**. Double-prime an explosive charge (at least 1/2 pound). Place the charge on top of the sight. Detonate the explosive electrically or nonelectrically.

b. **Gunfire**. Well-aimed shots from rifles or other small arms can make the sight useless to the enemy. When using small-arms fire, aim for critical parts such as the lenses, trigger assembly, and the electronics package.

c. **Mechanical Means**. Smash the sight with axes, picks, crowbars, rocks, and so on, or drive over it with a tracked vehicle. If you have enough time and personnel, destroy the sight completely.

d. **Fire**. After doing as much damage with the other methods, use vehicle fuels and lubricants to burn whatever is left of the sight.

### H-5. DESTRUCTION OF ROUND

When you destroy the missile, use the following methods:

| WARNING |
| --- |
| ***DO NOT*** **use mechanical methods to destroy live missiles.** |

a. **Launching**. The easiest and best way to destroy the missile is to fire it into enemy territory. After doing this, smash the launcher with axes, picks, and so on, or drive over it with a tracked vehicle.

b. **Exploding**. Explosives not only destroy the missile, but also the launcher. Placing one-half pound of explosive material on the round near the warhead should destroy the round (Figure H-1, page H-3).

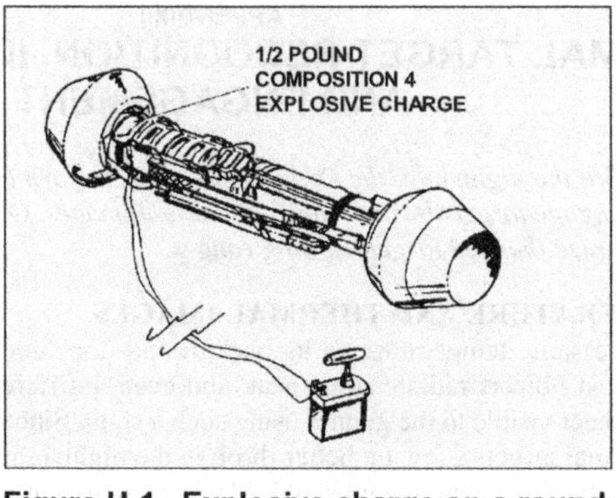


**Figure H-1. Explosive charge on a round.**

(1) Determine whether you can use *electric* blasting caps and wire or *nonelectric* blasting caps and safety fuzes to prime and detonate the explosive charges.

(a) If you prime the charges with electric blasting caps, take cover before firing them.

(b) If you prime the charges with nonelectric caps, crimp them onto at length of safety fuze at least 2 meters (6 1/2 feet) long. Once you prime the charges, ignite the safety fuzes and *take cover at once*.

(2) Use detonating cord to connect the charges. This produces a simultaneous detonation. Dual-prime the charges to reduce the possibility of a misfire.

c. **Burning**. If time allows after you have done as much damage as you can to the components using the other methods, burn the debris. Use vehicle fuels and lubricants to aid burning. You can also place one or more incendiary grenades on each component or part.

**DANGER**

1. **THE SAFETY FUZE BURNS AT THE RATE OF 1 FOOT IN 30 TO 40 SECONDS. BECAUSE SAFETY FUZE CONTAINS BLACK POWER AND BLASTING CAPS, YOU MUST PROTECT IT FROM MOISTURE AT ALL TIMES.**
2. **KEEP THE BLASTING CAPS, DETONATING CORD, AND SAFETY FUZES APART UNTIL YOU NEED TO USE THEM.**
3. **DO NOT FIRE SMALL ARMS AT THE MISSILE, AND KEEP OTHERS FROM DOING SO AS WELL. HITTING A LIVE ROCKET MOTOR OR A HIGH-EXPLOSIVE ANTITANK WARHEAD COULD INJURE OR KILL THE FIRER OR ANYONE ELSE NEAR THE MISSILE.**

APPENDIX I
# THERMAL TARGET RECOGNITION, IDENTIFICATION, AND ENGAGEMENT

*With the nightsight, the Dragon gunner can learn to recognize, identify, and engage targets by their unusual thermal images. On clear nights, he can recognize thermal targets at long ranges.*

## I-1. TEMPERATURE AND THERMAL IMAGES

An object the same temperature as its background does not show through a nightsight. However, most objects radiate some heat, and even a difference of less than one degree renders an object visible to the gunner using such a sight. Since temperatures generally drop at night, thermal targets show up better through the nightsight then.

a. Some targets, such as tanks and APCs, radiate heat in patterns recognizable to a trained, experienced Dragon gunner.

b. A soldier looking through a nightsight will not see an object of the same temperature as its background. However, the same soldier looking through the same sight will see as bright red any object or area on an object of greater heat than its background, such as engines that have recently been operated. Objects much colder than their backgrounds look black. Objects with less extreme temperatures appear as a less intense red or gray.

## I-2. SOURCES OF INFRARED ENERGY

The four sources of infrared energy include the sun, fuel combustion, friction, and radiant heat.

a. **Solar Heat**. Heat from the sun affects only the outside surfaces of objects.

(1) Solar heat highlights the outlines of an object, which helps the gunner recognize targets. For example, a solar-heated M113 APC looks like a box with a sloping front. However, a solar-heated M60A1 tank looks like a small oval on top of a larger oval. Gunners can detect objects at long ranges (out to 2,000 meters), but at medium ranges (800 to 1,200 meters), they can recognize specific shape cues. Profiles are usually easier to recognize than front views.

(2) Not only do atmospheric variables and surface reflections affect the color and intensity of a thermal image, so also does an object's ability to absorb sunlight. For example, dark-colored objects normally absorb sunlight better than light-colored objects.

b. **Fuel Combustion Heat**. An operating engine produces heat that warms up the surface of its surrounding compartment. Engine temperatures can reach up to 200 degrees Fahrenheit.

(1) ***Outside Features***. A gunner can seldom clearly see the actual shape of an engine, heated personnel space, muffler, or exhaust pipes, but the transferred heat may reveal features on the surface of the engine compartment.

(2) ***Relative Locations***. From the relative locations of what he perceives to be the engine and the exhaust, a trained and experienced gunner can tell whether the vehicle has a front- or rear-mounted engine.

(3) ***Directional Movement***. If the vehicle makes evasive maneuvers, the gunner can locate the vehicle's exhaust, which provides an important cue.

c. **Frictional Heat**. Moving parts generate frictional heat, which produces less intense heat than fuel combustion heat. The transport systems of moving vehicles generate frictional heat, which appears as a dim red in the nightsight. Except when high vegetation or dust obscures its transport system, the frictional heat it generates identifies a vehicle as wheeled or tracked.

(1) ***Tracked Vehicles***. In a moving tracked vehicle, the tracks, road wheels, drive sprockets, support rollers, and shock absorbers generate heat. When shock absorbers radiate the heat they have absorbed, the gunner can identify them at longer ranges. At short to medium ranges, radiant heat from shock absorbers could help the gunner identify the vehicle.

(2) ***Wheeled Vehicles***. In a moving wheeled vehicle, the tires, shock absorbers, differentials, drive shafts, transmissions, and axles generate heat. Gunners can detect the tires, shock absorbers, and differentials of a wheeled vehicle at medium to long range.

d. **Radiant Heat**. Smooth, shiny surfaces, such as windshields and glossy, painted fenders, reflect radiant heat from other sources. These reflections can produce odd images. For example, the fenders of an enemy tank appear black due to this thermal reflection; a glossy, painted vehicle could reflect off the tank's flat surfaces. An overcast sky can cause warmer thermal reflections. Diffuse surface reflections seldom cause problems.

## I-3. EFFECTS OF WEATHER AND OBSCURANTS

Variations in solar heat, fuel combustion heat, frictional heat, and thermal reflection affect infrared signatures and target recognition cues. Also, various atmospheric conditions affect the view through the nightsight, for better or worse.

a. **Precipitation**. Infrared energy transmits poorly through falling precipitation such as rain, snow, or fog, even when basic signature cues do not change. Falling rain and snow restrict visibility more than fallen rain and snow. During rain or snow, background objects and target features heated by friction and the sun lose their heat. Water, ice, or mud buildup on the transport system reduces frictional heat. The temperatures of engine compartments and exhausts remain high. Landmarks, such as trees, trails, and contour features, and background objects, such as trees and rocks, cool so much they may no longer be visible. This reduces scene clutter and can increase *target detection* capability. However, *target recognition* lessens as certain target features cool down. Rain and snow cool the target and reduce its contrast, causing a "snowy" image. The gunner adjusts the contrast control on the nightsight to compensate.

b. **Fallen Snow.** Snow reduces the temperatures of targets and background objects. Fallen snow evens out ground temperatures, which renders terrain features invisible. Without terrain features to use as a size reference, depth perception by size comparison becomes difficult.

c. **Dust, Diesel Fog, and Oil Smoke**. Dust particles created by the impact of artillery rounds reduce the nightsight's effectiveness.

## I-4. TARGET IDENTIFICATION

Vehicles have distinguishing characteristics or cues by which they can be classified and identified. Changes in atmospheric and ground conditions affect identification. To help compensate for these variables, the gunner must know how to use the control settings on the nightsight.

a. **Contrast and Brightness Controls**. Gunners can set the contrast and brightness (image) controls for the greatest internal detail of the target, then reset them to enhance other target-recognition cues. To increase their understanding of how these cues affect the thermal image and target recognition cues, gunners should experiment with the contrast and brightness controls (Figure I-1).

| BRIGHTNESS | CONTRAST | APPLICATION |
|---|---|---|
| Low | Medium to high | • Suppresses background clutter.<br>• Use in rainy, dusty, and lightly foggy conditions.<br>• Reveals hot objects only—cool ones will be invisible.<br>• When possible target appears, increase brightness and lower contrast gradually to reveal more thermal detail. |
| Medium to low | Medium | • Best overall setting for long ranges.<br>• Gradually reduce brightness to reveal cooler parts, such as tracks and road wheels, which will darken before the hotter parts do.<br>• Observe carefully while adjusting from medium to low and back. Gradual changes in brightness reveal distinctive features and aid in determining the hottest ones. |
| Medium | Medium to high | • Ranges greater than 1,200 meters.<br>• Best combination of settings to reveal the hull shapes and overall silhouettes of target vehicles.<br>• Especially useful for—<br>– Identifying small target images with few recognizable features.<br>– Tracking long-range targets in poor visibility such as heavy fog or dust.<br>– Searching tree lines in wet conditions. Limitation of this setting: a snowy, distorted, tough-to-interpret image does not enhance target identification. |

**Figure I-1. Setting combinations for contrast and brightness controls.**

b. **Focus Controls**. Most nightsight focus controls are sensitive; that is, a small movement of the control knob causes a large change in focal point. If a gunner has trouble focusing, he should check the image (contrast and brightness) controls.

(1) To adjust the reticle, turn the focus control (located on the eyepiece).

(2) Adjust the range focus lever. Before focusing the objective lens, move the image controls (contrast and brightness) to the low-to-medium level. Gunners learn to focus the objective lens skillfully through trial and error. This presents a challenge, because infrared heat diffuses on objects. It seldom yields the clear, straight lines that other types of sights do. Thus, you must learn to focus by moving the control back and forth to obtain the best image. Start by aiming the nightsight at a prominent object. Once you have determined the best image, you can experiment by focusing on different objects at different ranges.

## I-5. BATTLEFIELD IDENTIFICATION

Using a nightsight to identify targets on a battlefield presents a challenge also. In a target-rich environment on a dry, clear night, high-confidence identification requires a thermal image of such features as road wheels, turret shapes, gun tube, and exhaust location. In clear weather, a gunner can thermally distinguish an M60A1 main battle tank from a T-62 medium tank at a range of 800 to 1,200 meters. To identify a target, a gunner should ask himself the following questions:

- Is the target moving? In what direction?
- Where is the engine? Where is the exhaust?
- Is the target in the unit's section? Should it be there?
- Is the target in a formation?
- Is the target firing at friendly or enemy units?

## I-6. PRIMARY RECOGNITION CUES

The friendly and enemy cues shown in the following figures help in training gunners to recognize vehicles.

- Figure I-2. M60A1 main battle tank (page I-5).
- Figure I-3. M551 light tank (page I-6).
- Figure I-4. M113 armored personnel carrier (page I-7).
- Figure I-5. Trucks (page I-8).
- Figure I-6. T-62 medium tank (page I-9).
- Figure I-7. BMP infantry combat vehicle (page I-10).
- Figure I-8. BTR-60 armored personnel carrier (page I-11).
- Figure I-9. BRDM-2 reconnaissance vehicle (page I-12).

## M60A1 Main Battle Tank

***Classification***. Main characteristics of the M60A1 main battle tank include—

- A rear engine.
- An oval-shaped track and road wheel pattern.
- An overall hull, turret, and gun pattern that may be visible with the maximum setting.
- A gun tube that can be seen when the gun has recently fired.

***Side-View Identification***. M60A1 main battle tank characteristics visible from both sides include—

- A rear engine and a rearward exhaust.
- A high-profile track pattern with hot, taut tracks. (The gunner may see the six evenly spaced road wheels and three support rollers, especially at short ranges.)
- A large, centrally mounted turret.
- A medium-length gun tube that can be seen at short ranges. When the gun has recently fired, the gun tube can be seen at long ranges. (The bore evacuator is two-thirds of the way down the length of the barrel.)
- A high overall profile with a large turret mounted in the center.

***Front View Identification***. M60A1 main battle tank characteristics visible from the front include—

- Two warm tracks, separated by a cool hull. The lower front hull will appear warm if the personnel heater has operated recently.
- A cool front hull, which indicates a rear-engine vehicle.
- One track that seems wider, if the vehicle sits slightly oblique to the viewer.
- A gun tube that observers can see when the gun has recently fired.

***Effects of Motion***. The M60A1 main battle tank's direction of movement may indicate the location of its engine. Other characteristics that observers could see when the M60A1 main battle tank moves include—

- A transport system that becomes warmer and more visible during movement.
- A slower bounce than a lighter vehicle.
- A sometimes visible exhaust plume, whose direction indicates the direction of movement.

***Distinguishing Features***. The side view of the M60A1 track pattern appears higher and more uniformly warm than the side view of the T-62. The tracks of the T-62 are cooler toward the front and smaller than those of the M60A1 tank. From the front, the M60A1 main battle tank appears hotter and larger than the T-62 medium tank.

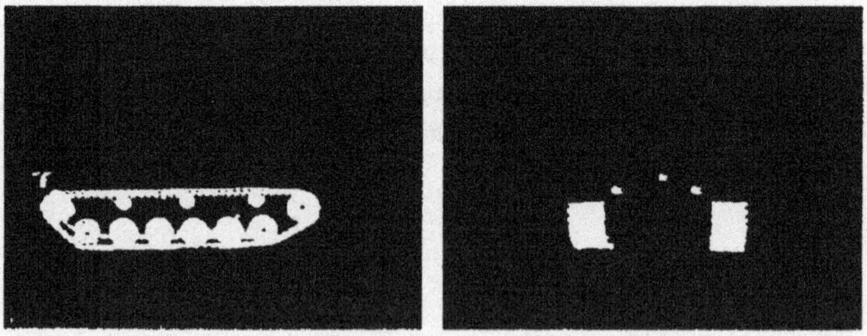

**Figure I-2. M60A1 main battle tank.**

## M551 Light Tank

***Classification***. Main characteristics of the M551 light tank include—

- A rear engine.
- An oval-shaped track and road wheel pattern.
- A combined hull, turret, and gun pattern.

***Side-View Identification***. M551 light tank characteristics visible from both sides include—

- A rear engine and rear exhaust, whose plume may be directed upward or rearward.
- A low profile: low hull and a small, low, flat turret.
- A wedge-like shape, with the wedge pointing to the vehicle's front and visible between 800 and 1,200 meters.
- A high side decking.
- Warm, slack tracks and (at short ranges) five evenly spaced road wheels.
- A short gun tube that can be seen when the gun has recently fired.

***Front-View Identification***. M551 light tank characteristics visible from the front include—

- A wide, low turret, whose sides extend almost over the tracks, providing a unique front view.
- Warm tracks separated by a cool hull, visible as two red spots.

***Effects of Motion***. The M551 light tank's direction of movement may indicate the location of its engine. Other characteristics that observers could see when the M551 light tank moves include—

- A transport system that becomes warmer and more visible during movement.
- A slower bounce than a lighter vehicle.

***Distinguishing Features***. The M551 light tank has a wide, low turret (front view) and a wedge shape (side view). Otherwise, its signatures are difficult to distinguish from those of a T-62 tank.

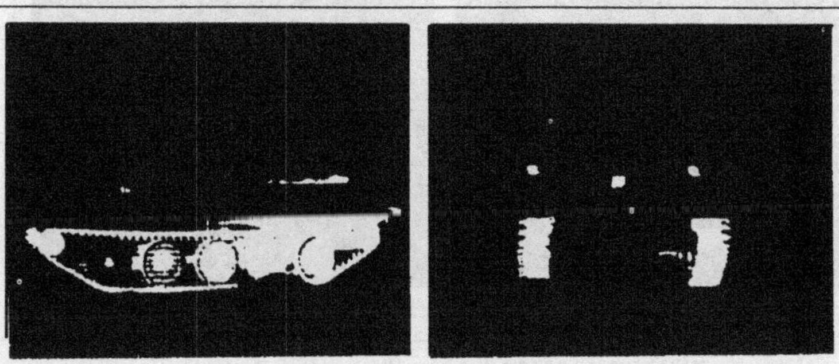

**Figure I-3. M551 light tank.**

## M113 Armored Personnel Carrier

***Classification***. Main characteristics of the front-engine M113 APC include—

- An oval-shaped track and road wheel pattern.
- A box-shaped hull with no turret.

***Side-View Identification***. From the right, even at long ranges, you can see the M113 APC's characteristic front hot corner. From the left, you can see a cool box shape. Characteristics visible from both sides include—

- A beveled front end.
- A low track.
- A track skirt, which gives the track a low profile.
- Five evenly spaced road wheels.

***Front-View Identification***. The M113 APC's characteristic hot spot can be seen on its front right corner. Other characteristics visible from the front include—

- A dark surfboard across the front of the vehicle.
- An overall boxy shape (at the maximum range setting).
- On the front right corner, an upward-moving exhaust plume that can be seen at short ranges.

***Effects of Motion***. The M113 APC's direction of movement may indicate the location of its engine, which will get hotter and more visible as it operates. Other characteristics that observers could see when the M113 APC moves include—

- A transport system that becomes warmer and more visible during movement.
- A slower bounce than a lighter vehicle, but a faster bounce than a tank.
- A sometimes visible exhaust plume whose direction indicates the direction of movement.

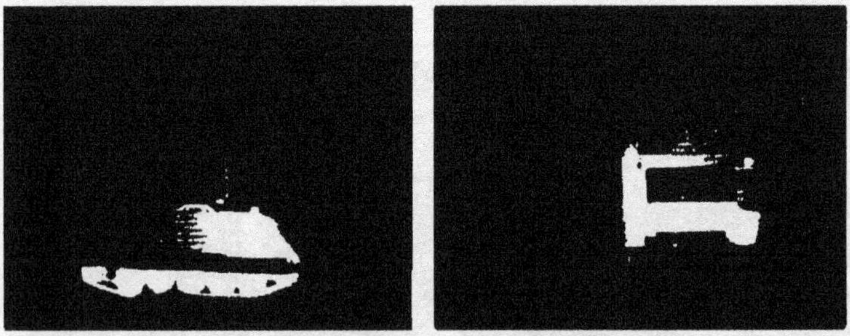

**Figure I-4. M113 armored personnel carrier.**

## Trucks

**(All truck cues are combined)**

External size cues are needed to establish the size of a truck. To estimate the size of a cue, the gunner compares a nearby tree or other background feature to the size of the red spot. He can also use the reticle to gage size. This is most accurate at low to medium settings.

***Side-View Identification***. Truck characteristics visible from both sides include—

- A prominent square, also visible from the front, that indicates a front engine. The entire hood and radiator usually appear hot, and the front wheels may seem to merge into the image of the engine.
- Warm spots, separated from the front wheels, that indicate rear wheels. These are easily distinguished from a track pattern.
- Silhouettes of the cab and hood, which may appear at maximum settings. (Long-bed trucks often appear as other truck shapes from the side, even at long ranges.)
- A diagonally-angled drive shaft that is often visible as a bright area joining the engine and rear wheels.
- Exhaust pipes and stacks, which appear hot along their entire length. These appear as different shapes, depending upon how they are routed through the vehicle's frame.

***Front-View Identification***. Truck characteristics commonly visible from the front include—

- A bright red square or rectangle, indicating the engine compartment, hood, radiator, and, if a personnel heater is in use, the cab. Front views of trucks are all similar at medium and long ranges.
- Wheels and a front axle area that often merge into a continuous red spot beneath the engine. Lower settings improve this image so the gunner can see the vehicle's wheels.
- Canvas and wood parts. These parts usually appear cool at night and warm during the day.

**Figure I-5. Trucks.**

## T-62 Medium Tank

***Classification***. Main characteristics of the T-62 medium tank include—

- A rear engine.
- An oval-shaped track and road wheel pattern.
- A combined hull, turret, and gun pattern that may be visible at maximum range setting.
- A gun tube that can be seen when the gun has recently fired.

***Side-View Identification***. On the left side of the T-62 medium tank, a large heated area and exhaust flume indicate a rear engine. T-62 medium tank characteristics visible from both sides include—

- Low-profile, slack tracks and, at short ranges, five road wheels with unique (unequal) spacing.
- A low overall profile with a cool hull.
- A long gun tube that can be seen when the gun has recently fired.
- A small, centrally mounted turret.

***Front-View Identification***. T-62 medium tank characteristics visible from the front include—

- A cool, low overall profile, with a left side that is warmer than the right.
- Warm tracks separated by a cool hull and cool fenders that may appear black above the tracks.
- A small, dome-shaped turret.
- A long gun tube that can be seen when the gun has recently fired. When aimed straight at the viewer, it appears as a red spot. Its firing signature can be seen only at short ranges.

***Effects of Motion***. The direction of movement may reveal the location of the engine. Other characteristics that observers could see when the T-62 medium tank moves include—

- A transport system that becomes warmer and more visible during movement.
- A slower bounce than a lighter vehicle.
- A sometimes visible exhaust plume whose direction indicates the direction of movement.

**Figure I-6. T-62 medium tank.**

## BMP Infantry Combat Vehicle

***Classification***. Main characteristics of the BMP infantry combat vehicle include—

- An engine mounted in the front right of the vehicle and visible from its front and right.
- A taut track pattern and, at close ranges, visible wheels and support rollers.
- A low overall profile that can be seen at higher settings.
- A gun tube that can be seen when the gun has recently fired.

***Side-View Identification***. The right side of the BMP infantry combat vehicle is hotter than its left side and is usually more recognizable than other views. The engine appears as a large red spot in the forward half, and the exhaust port is hotter and brighter red than the engine. The tracks and engine area appear together to form a boat-like shape on the right side. BMP characteristics visible from both sides include—

- A track pattern that can be seen at long ranges.
- A forward end that slopes upward from the track to the front fender.
- Cool track fenders that block the view of the upper track.
- A long, low shape that can be seen at high settings.

***Front-View Identification***. BMP characteristics visible from the front include—

- An engine located in the right front, indicated by a square-shaped red spot.
- A cool, front deck panel, indicated by a black line across the engine's red spot.
- A sometimes visible exhaust plume which vents upward from the right side of the vehicle.
- An exhaust port that can be seen as a small red spot in the top-right corner of the engine's larger red spot.
- Visible tracks, with a small gap separating the red spot of the vehicle's left track from the engine's red spot.
- A turret that cannot be seen at long ranges.

***Effects of Motion***. The direction of movement may reveal the location of the engine, and changing target views may reveal other features such as—

- A transport system that becomes warmer and more visible during movement.
- A slower bounce than a lighter vehicle.
- A sometimes visible exhaust plume whose direction may indicate the direction of movement.

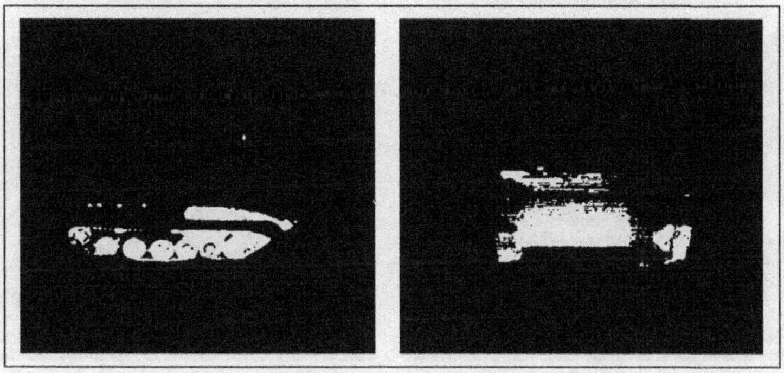

**Figure I-7. BMP infantry combat vehicle.**

## BTR-60 Armored Personnel Carrier

***Classification.*** Main characteristics of the BTR-60 include—

- A rear engine and multiple wheels that can be distinguished from the side at long ranges.
- Mufflers on the rear deck that intensify the rear-engine compartment cue.
- A long, high profile, apparent even at long ranges.
- A gun tube that can be seen when the gun has recently fired.

***Side-View Identification.*** BTR-60 characteristics visible from both sides include—

- A rear engine, which creates the brightest spot on the image.
- Four large, inflatable tires on each side, with a gap between the front and rear pairs.
- Two mufflers mounted over the rear engine compartment.
- A small turret that can be seen at medium ranges and that sometimes can be seen at long ranges.
- Identical left-side and right-side views.

***Front-View Identification.*** BTR-60 characteristics visible from the front include—

- A pattern across the front surface, caused by variations in temperature, visible as multiple shapes.
- A rounded hull.
- A cool surfboard that divides the top and bottom halves of the front hull.
- A cool front, unless the personnel heater is in use.
- Shock absorbers that can be seen at medium ranges.

***Effects of Motion.*** The vehicle's direction of movement may indicate the location of its engine, which will get hotter as it operates and thus brighter red. Other characteristics that observers could see when the BTR-60 moves include—

- Its wheels, which will become more visible as they get warmer.
- Its mufflers, which will also become more visible as they get hot.
- A slow bounce, even on rough terrain, which identifies it as a large vehicle.
- A sometimes visible exhaust plume whose direction indicates the direction of movement.

**NOTE:** Some of the heat from the muffler merges with the heat from the front hull and wheels, making a front three-quarter view of this vehicle more confusing than the same view of other vehicles.

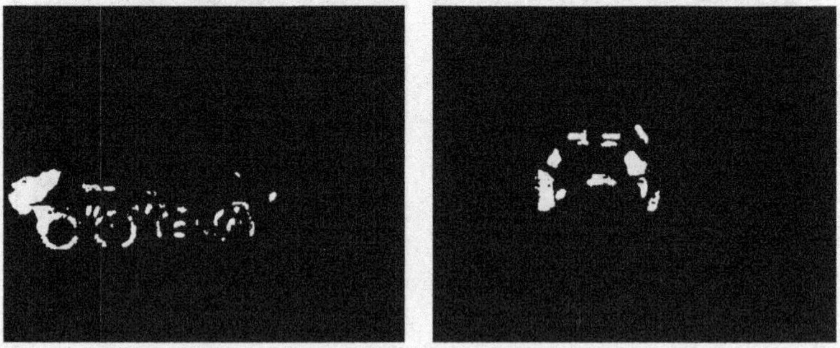

**Figure I-8. BTR-60 armored personnel carrier.**

## BRDM-2 Reconnaissance Vehicle

***Classification***. Main characteristics of the BRDM-2 reconnaissance vehicle include—

- A rear engine and exhaust.
- A pattern of separate, wheel-sized red spots.
- An overall profile of a hull with (or without) turret.
- A gun tube that can be seen when the gun has recently fired.

***Side-View Identification***. BRDM-2 characteristics visible from both sides include —

- The vehicle's small overall size, indicated in part by the short distance between the two tires clearly visible on each side.
- A distinctive heat pattern formed by the rear mufflers and wheels. This pattern is clear even at long ranges. Left and right views are the same.
- The rear engine, which is clearly visible at long ranges.
- Two mufflers, one on each side of the top rear deck, which enlarge the hot spot created by the engine area.
- A choppy ride over rough terrain that is noticeable even at long ranges.

***Front-View Identification***. BRDM-2 characteristics visible from the front include —

- Two warm tires separated by a cool hull. The red spot of the differential may be visible between the front tires.
- A cool front hull, indicating a rear engine.

**NOTE:** Setting the brightness and contrast controls on high makes the entire front of this vehicle appear hot.

- The dark surfboard across the front of the hull can be seen.
- The tires are set closer together than the tracks of most armored vehicles.
- A greater height-to-width ratio than most armored vehicles, so that it appears narrower and taller.
- Over rough terrain, the vehicle tends to bounce more than armored vehicles.

***Effects of Motion***. The vehicle's direction of movement may indicate the location of its engine. Other characteristics that observers could see when the BRDM-2 reconnaissance vehicle moves include—

- A transport system that becomes warmer and more visible during movement.
- A sometimes visible exhaust plume whose direction indicates the direction of movement.

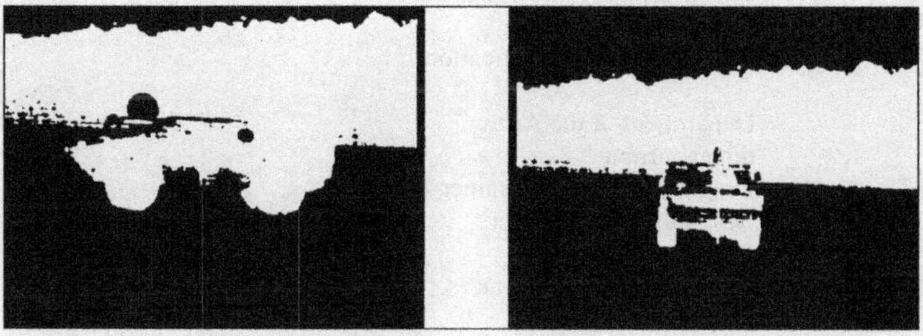

**Figure I-9. BRDM-2 reconnaissance vehicle.**

# GLOSSARY

## ACRONYMS AND ABBREVIATIONS

| | |
|---|---|
| AAR | after-action review |
| amp | ampere |
| APC | armored personnel carrier |
| AR | Army regulation |
| ARTEP | Army Training and Evaluation Program |
| ASI | additional skill identifier |
| asst | assistant |
| ATGM | antitank guided missile |
| ATWESS | antitank weapon effect signature simulator |
| batt | battery |
| BFV | Bradley fighting vehicle |
| BMP | (a fighting vehicle manufactured by the former Soviet Union) |
| BRDM | (a scout car manufactured by the former Soviet Union) |
| BRT | brightness |
| BTR | (a personnel carrier manufactured by the former Soviet Union) |
| C2 | command and control |
| CB | citizen's band (radio) |
| cbt | combat |
| CFV | cavalry fighting vehicle |
| CP | command post |
| CPX | command post exercise |
| CS | combat support |
| CTA | common table of allowances |
| CTRS | contrast |
| CVI | combat vehicle identification |
| DA | Department of the Army |
| DC | direct current |
| DFTT | Dragon field tactical trainer |
| DGT | Dragon gunnery trainer |
| dir | direction |
| DLIC | detachment left in contact |
| DZ | drop zone |
| EOD | explosive ordnance disposal |
| F | Fahrenheit |
| 1SG | first sergeant |
| FASCAM | family of scatterable mines |

| | |
|---|---|
| FHT | field handling trainer |
| FIST | fire support team |
| FM | field manual |
| FOV | field of view |
| FPF | final protective fire |
| FPL | final protective line |
| freq | frequency |
| FSO | fire support officer |
| FTX | field training exercise |
| GTA | graphic training aid |
| HE | high explosive |
| headphone | An earphone held over the ear by a band worn on the head—usually used in plural, as in "*a set of headphones*." |
| headset | An attachment for holding an earphone and transmitter at one's head. |
| HEAT | high-explosive antitank |
| HMMWV | high-mobility multipurpose wheeled vehicle |
| horz | horizontal |
| HQ | headquarters |
| hr | hour |
| Hz | Hertz |
| IAW | in accordance with |
| KHz | kilohertz |
| KPH | kilometers per hour |
| LAW | light antitank weapon |
| LCE | load-carrying equipment |
| LD | line of departure |
| ldr | leader |
| LFX | live-fire exercise |
| LOS | line of sight |
| LP | listening post |
| L-R | left to right |
| LRA | local reproduction authorized |
| LTID | laser target interface device |
| M | monthly |
| mech | mechanized |
| METT-T | mission, enemy, terrain (and weather), troops, and time available |
| MILES | multiple integrated laser-engagement system |
| min | minute(s) |
| mm | millimeter |

| | |
|---|---|
| mod | model |
| mon | monitor |
| MOPP | mission-oriented protective posture |
| MOUT | military operations on urbanized terrain |
| MPH | miles per hour |
| MQS | military qualification standards |
| MTP | mission training plan |
| | |
| NA | not applicable |
| NATO | North Atlantic Treaty Organization |
| NAVTRADEV | Naval training development |
| NBC | nuclear, biological, and chemical |
| NCO | noncommissioned officer |
| NCOES | Noncommissioned Officer Education System |
| NCOIC | noncommissioned officer in charge |
| nonmech | nonmechanized |
| NVD | night vision device |
| | |
| OPFOR | opposing force |
| OSUT | one-station unit training |
| | |
| PDF | principal direction of fire |
| PGTS | Precision Gunnery Training System |
| plt | platoon |
| PMCS | preventive maintenance checks and services |
| POI | program of instruction |
| PSG | platoon sergeant |
| PSI | pounds per square inch |
| PSM | power supply modulator |
| pwr | power |
| | |
| Q | quarterly |
| qual | qualification |
| qtr | quarterly |
| | |
| RDL | Reimer Digital Library |
| RSTA | reconnaissance, surveillance, and target acquisition |
| | |
| S2 | intelligence officer |
| S3 | operations and training officer |
| sec | seconds |
| SGT | sergeant |
| SL | skill level |
| SMCT | soldier's manual of common tasks |
| SOP | standing operating procedure |

| | |
|---|---|
| sqd | squad |
| SSN | social security number |
| STB | supertropical bleach |
| STP | soldier's training publication |
| STX | situational training exercise |
| TC | training circular |
| tm | team |
| TM | technical manual |
| tng | training |
| TO | training objective |
| TOW | tube-launched, optically tracked, wire-guided (missile) |
| TRADOC | (United States Army) Training and Doctrine Command |
| TRP | target reference point |
| TSC | Training Support Center |
| TTP | tactics, techniques, and procedures |
| US | United States |
| USC | United States code |
| VAC | voltage alternating current |
| VDC | voltage direct current |
| vert | vertical |
| VPC | vehicle power conditioner |
| WP | white phosphorus |
| XO | executive officer |

# REFERENCES

## SOURCES USED

These are the sources quoted or paraphrased in this publication:

| | |
|---|---|
| AR 385-62 | Regulations for Firing Guided Missiles and Heavy Rockets for Training, Target Practice and Combat. 5 January 1977; with Change 1, 1 Jun 1983. |
| AR 385-63 | Policies and Procedures for Firing Ammunition for Training, Target Practice, and Combat. 15 October 1983. |
| AR 614-200 | Enlisted Assignments and Utilization Management. 20 November 2000. |
| FM 3-5 | NBC Decontamination. 28 July 2000. |
| FM 7-7 | The Mechanized Infantry Platoon and Squad (APC). 15 March 1985. |
| FM 7-7J | The Mechanized Infantry Platoon and Squad (Bradley). 7 May 1993. |
| FM 7-8 | Infantry Rifle Platoon and Squad (Infantry, Airborne, Air Assault, Ranger). 22 April 1992. |
| FM 7-10 | The Infantry Rifle Company. 14 December 1990. |
| FM 21-11 | First Aid for Soldiers. 27 October 1988; with Change 1, 28 August 1989 and Change 2, 4 December, 1991. |
| FM 71-1 | Tank and Mechanized Company Team. 26 January 1998. |
| STP 21-1-SMCT | Soldier's Manual of Common Tasks, Skill Level 1. 1 October 1994. |
| STP 21-24-SMCT | Soldier's Manual of Common Tasks (SMCT), Skill Levels 2/3/4. 1 October 1992. |
| TM 9-1265-368-10-1 | Operator's Manual for Multiple Integrated Laser Engagement System (MILES) Simulator System, Firing Laser: M62 for Dragon Weapons System. 15 July 1988; with Change 1, 16 August 1989. |

TM 9-1425-484-10 Operator's Manual for Dragon Weapon Guided Missile System, Surface Attack: M47 (Medium Antitank/Assault Weapon System). 31 July 1979; with Changes 1 thru 13, 17 October 1980 thru 30 March 1993.

TM 9-5855-254-14-HR Hand Receipt Covering Content of Components of End Item (COEI) Basic Issue Items (BII), and Additional Authorization List (AAL) for Charger, Battery PP-7382/TAS. 22 October 1987.

**DOCUMENTS NEEDED**

These documents must be available to the intended users of this publication:

*Publications*

ARTEP 7-8-MTP Mission Training Plan for the Infantry Rifle Platoon and Squad. 3 March 2000 (RDL version).

GTA 17-2-11 Combat Vehicle Identification Training Cards (CVI). 1984.

GTA 17-2-13 Armored Vehicle Recognition. 1987.

TB Med 524 Occupational and Environmental Health: Control of Hazards to Health from Laser Radiation. 20 June 1985.

*Forms*

The forms prescribed by this manual can be obtained from three sources.

- *Reproducible Forms*--Copy the blank forms from the back of the manual onto 8 1/2-by 11-inch paper.
- *CD-ROM*--Find the forms on the Army Electronic Library (AEL) disks (EM0001).
- *Downloadable Forms*--Download the blank forms in PDF, HTML, BOO, or FRO from the USAPA website (http://www.usapa.army.mil).
- *Electronic Forms*--Complete FRZ and FRO forms electronically (except for signature blocks, which must be completed manually) using *FormFlow Filler*. Download the latest version of FormFlow Filler software from the same website. NOTE: If any item on the blank form prints in the wrong place, turn Enhanced Fonts on or off in the Tools, Options menu.

DA Form 4242-1-R Dragon Performance Scorecard.

DA Form 7445-R Dragon Gunnery Trainer (DGT) Scorecard, Firing Tables 1-4.

| | |
|---|---|
| DA Form 5517-R | Standard Range Card (LRA). February 1986. (For a locally reproducible copy of this form, see FM 7-7J.) |
| DA Form 7446-R | Dragon Field Tactical Trainer (DFTT) Scorecard, Firing Tables 5-6. |
| DA FORM 7447-R | Dragon Gunnery Trainer (DGT) Scorecard, Firing Tables 7-8. |

*Internet Web Sites*

U.S. Army Publishing Agency (USAPA)
http://www.usapa.army.mil

Army Doctrine and Training Digital Library (ADTDL)
http://www.adtdl.army.mil

Air Force Publishing
http://afpubs.hq.af.mil/

**READINGS RECOMMENDED**

These sources contain relevant supplemental information:

| | |
|---|---|
| AR 614-200 | Enlisted Assignments and Utilization Management. 20 November 2000. |
| CTA 50-970 | Expendable/Durable Items (Except Medical, Class V, Repair Parts, and Heraldic Items), 21 September 1990. |
| DA Label 80 | US Army Calibrated Instrument, 1 December 1977. |

# INDEX

# DRAGON PERFORMANCE SCORECARD

For use of this form see FM 3-23.24 (FM 23-24); the proponent agency is TRADOC.

DATA REQUIRED BY PRIVACY ACT OF 1974. AUTHORITY: 10 USC 3012(g)/Executive Order 9397. PRINCIPLE PURPOSE(S): Aids in individual training on targets at various ranges. ROUTINE USE(S): To evaluate individual proficiency. SSN used for positive identification purposes only. MANDATORY OR VOLUNTARY DISCLOSURE AND EFFECT ON INDIVIDUAL NOT PROVIDING INFORMATION: Voluntary. Individuals not providing information cannot be rated or scored on a mass basis.

| LAST NAME | FIRST NAME | MI |
|---|---|---|
| | | |

| UNIT | SSN | RANK |
|---|---|---|
| | | |

| | TEST DATE (YYYYMMDD) | | RETEST DATE (YYYYMMDD) | | RETEST DATE (YYYYMMDD) | |
|---|---|---|---|---|---|---|
| | GO | NO-GO | GO | NO-GO | GO | NO-GO |
| 1. Maintain the M47 medium antitank weapon. | | | | | | |
| 2. Prepare the M47 for firing. | | | | | | |
| 3. Restore the M47 to carrying configuration. | | | | | | |
| 4. Demonstrate correct firing positions. | | | | | | |
| 5. Determine if targets are engageable. | | | | | | |
| 6. Perform immediate-action procedures. | | | | | | |
| 7. Prepare an antiarmor range card. | | | | | | |
| 8. Identify the six major components of an M47 firing position. | | | | | | |
| 9. Identify armored vehicles. | | | | | | |
| 10. Explain emergency destruction procedures. | | | | | | |
| 11. Explain emergency decontamination procedures. | | | | | | |
| **UNQUALIFIED** (INSTRUCTOR'S INITIALS) | | | | | | |
| **QUALIFIED** (INSTRUCTOR'S INITIALS) | | | | | | |
| INSTRUCTOR'S NAME (PRINT) | | | | | | |
| INSTRUCTOR'S SIGNATURE | | | | | | |

NOTES:

1. The instructor and Dragon gunners must pass all tasks.
2. A soldier eliminates himself from training if, after three tries -- the initial test and two retests -- he still fails to meet the performance standards on any task.
3. The instructor scores each test by initialing in the GO or NO-GO columns for that test.
4. After each test, the instructor initials below the GO-NO GO columns in either the "qualified" or "unqualified" block.
5. After the soldier either qualifies or fails to qualify after three tries, the instructor prints and signs his own name in the lower righthand corner of this scorecard.

REMARKS:

DA FORM 4242-1-R, DEC 2001

USAPA V1.00

# DRAGON GUNNERY TRAINER (DGT) SCORECARD

## MOVING AND STATIONARY TARGETS

For use of this form see FM 3-23.24 (FM 23-24); the proponent agency is TRADOC.

DATA REQUIRED BY PRIVACY ACT OF 1974. AUTHORITY: 10 USC 3012(g)/Executive Order 9397. PRINCIPLE PURPOSE(S): Aids in individual training on targets at various ranges. ROUTINE USE(S): To evaluate individual proficiency. SSN used for positive identification purposes only. MANDATORY OR VOLUNTARY DISCLOSURE AND EFFECT ON INDIVIDUAL NOT PROVIDING INFORMATION: Voluntary. Individuals not providing information cannot be rated or scored on a mass basis.

| LAST NAME | FIRST NAME | MI | SSN | UNIT NO. | POINT NO. | DATE (YYYYMMDD) |
|---|---|---|---|---|---|---|
| | | | | | | |

## FIRING TABLE 1

STATIONARY AND MOVING TARGETS
DISK: DGT #14
PERCENT HIT ZONE SETTING 50 75 100 %

EVENT A. OPTICAL SIGHT. WEIGHT LOSS OR PULL-DOWN; AIMING POINT AND TARGET ASPECT.

| ENG | MSN | OBSC | HIT | MISS | TRACKING SCORE | THRUSTERS |
|---|---|---|---|---|---|---|
| 1 | 1 | 0.0 | | | | |
| 2 | 2 | 0.0 | | | | |
| 3 | 3 | 0.5 | | | | |
| 4 | 3 | 1.0 | | | | |
| 5 | 4 | 2.0 | | | | |

EVENT B. OPTICAL SIGHT. INTRODUCTION TO OBSCURATION AND MOVING.

| ENG | MSN | OBSC | HIT | MISS | TRACKING SCORE | THRUSTERS |
|---|---|---|---|---|---|---|
| 1 | 8 | 0.5 | | | | |
| 2 | 9 | 1.0 | | | | |
| 3 | 10 | 0.5 | | | | |
| 4 | 11 | 0.5 | | | | |
| 5 | 12 | 0.5 | | | | |

## FIRING TABLE 2

STATIONARY AND MOVING TARGETS
DISK: DGT #14
PERCENT HIT ZONE SETTING 50 75 100 %

EVENT A. OPTICAL SIGHT. STATIONARY AND MOVING TARGETS.

| ENG | MSN | OBSC | HIT | MISS | TRACKING SCORE | THRUSTERS |
|---|---|---|---|---|---|---|
| 1 | 8 | 0.5 | | | | |
| 2 | 9 | 0.5 | | | | |
| 3 | 18 | 0.0 | | | | |
| 4 | 19 | 0.0 | | | | |
| 5 | 20 | 0.0 | | | | |

EVENT B. BRIGHTNESS AND CONTRAST.

| ENG | MSN | OBSC | HIT | MISS | TRACKING SCORE | THRUSTERS |
|---|---|---|---|---|---|---|
| 1 | 26 | 0.0 | | | | |
| 2 | 27 | 0.0 | | | | |
| 3 | 28 | 0.0 | | | | |
| 4 | 29 | 0.0 | | | | |
| 5 | 30 | 0.0 | | | | |

## FIRING TABLE 3

MOVING TARGETS
DISK: DGT #14
PERCENT HIT ZONE SETTING 50 75 100 %

EVENT A. OPTICAL SIGHT.

| ENG | MSN | OBSC | HIT | MISS | TRACKING SCORE | THRUSTERS |
|---|---|---|---|---|---|---|
| 1 | 8 | 1.0 | | | | |
| 2 | 9 | 1.0 | | | | |
| 3 | 10 | 1.0 | | | | |
| 4 | 11 | 1.0 | | | | |
| 5 | 12 | 1.0 | | | | |

EVENT B. THERMAL SIGHT.

| ENG | MSN | OBSC | HIT | MISS | TRACKING SCORE | THRUSTERS |
|---|---|---|---|---|---|---|
| 1 | 26 | 0.0 | | | | |
| 2 | 27 | 0.0 | | | | |
| 3 | 28 | 0.0 | | | | |
| 4 | 29 | 0.0 | | | | |
| 5 | 30 | 0.0 | | | | |

## FIRING TABLE 4

INTERMEDIATE TARGETS
DISK: DGT #14
PERCENT HIT ZONE SETTING 50 75 100 %

EVENT A. OPTICAL SIGHT.

| ENG | MSN | OBSC | HIT | MISS | TRACKING SCORE | THRUSTERS |
|---|---|---|---|---|---|---|
| 1 | 7 | 0.5 | | | | |
| 2 | 9 | 1.0 | | | | |
| 3 | 10 | 1.5 | | | | |
| 4 | 11 | 0.5 | | | | |
| 5 | 12 | 1.0 | | | | |

EVENT B. OPTICAL AND THERMAL SIGHT.

| ENG | MSN | OBSC | HIT | MISS | TRACKING SCORE | THRUSTERS |
|---|---|---|---|---|---|---|
| 1 | 13 | THERMAL | | | | |
| 2 | 26 | THERMAL | | | | |
| 3 | 28 | THERMAL | | | | |
| 4 | 29 | THERMAL | | | | |
| 5 | 30 | THERMAL | | | | |

EVENT C. OPTICAL SIGHTS, MOPP, PROTECTIVE MASK FAMILIARIZATION.

| ENG | MSN | OBSC | HIT | MISS | TRACKING SCORE | THRUSTERS |
|---|---|---|---|---|---|---|
| 1 | 7 | 0.0 | | | | |
| 2 | 9 | 0.0 | | | | |
| 3 | 10 | 0.0 | | | | |
| 4 | 11 | 0.0 | | | | |
| 5 | 12 | 0.0 | | | | |

LEGEND: ENG - ENGAGEMENT
MSN - MISSION
OBSC - OBSCURATION

REMARKS:

SCORER'S NAME (PRINT) ______________________

SCORER'S SIGNATURE ______________________

DA FORM 7445-R, DEC 2001

USAPA V1.00

# DRAGON FIELD TACTICAL TRAINER (DFTT) SCORECARD

## MULTIPLE TARGETS

**TARGET SIZE SETTING: 1 2 3**

For use of this form see FM 3-23.24 (FM 23-24); the proponent agency is TRADOC.

DATA REQUIRED BY PRIVACY ACT OF 1974. AUTHORITY: 10 USC 3012(g)/Executive Order 9397. PRINCIPLE PURPOSE(S): Aids in individual training on targets at various ranges. ROUTINE USE(S): To evaluate individual proficiency. SSN used for positive identification purposes only. MANDATORY OR VOLUNTARY DISCLOSURE AND EFFECT ON INDIVIDUAL NOT PROVIDING INFORMATION: Voluntary. Individuals not providing information cannot be rated or scored on a mass basis.

| LAST NAME | FIRST NAME | MI | SSN | UNIT NO. | POINT NO. | DATE (YYYYMMDD) |
|---|---|---|---|---|---|---|
| | | | | | | |

## FIRING TABLE 5

MULTIPLE TARGETS, FAMILIARIZATION ONLY

EVENT A. OPTICAL SIGHT, SITTING.

| TGT | RANGE | *OBS | HIT KILL | HIT NO KILL | MISS | MISSILE GROUNDED | EXCEEDED RANGE | THRUSTERS |
|---|---|---|---|---|---|---|---|---|
| 1 | | | | | | | | |
| 2 | | | | | | | | |
| 3 | | | | | | | | |
| 4 | | | | | | | | |
| 5 | | | | | | | | |

EVENT B. OPTICAL SIGHT, STANDING.

| TGT | RANGE | *OBS | HIT KILL | HIT NO KILL | MISS | MISSILE GROUNDED | EXCEEDED RANGE | THRUSTERS |
|---|---|---|---|---|---|---|---|---|
| 1 | | | | | | | | |
| 2 | | | | | | | | |
| 3 | | | | | | | | |
| 4 | | | | | | | | |
| 5 | | | | | | | | |

EVENT C. THERMAL SIGHT, SITTING.

| TGT | RANGE | *OBS | HIT KILL | HIT NO KILL | MISS | MISSILE GROUNDED | EXCEEDED RANGE | THRUSTERS |
|---|---|---|---|---|---|---|---|---|
| 1 | | | | | | | | |
| 2 | | | | | | | | |
| 3 | | | | | | | | |
| 4 | | | | | | | | |
| 5 | | | | | | | | |

EVENT D. THERMAL SIGHT, STANDING.

| TGT | RANGE | *OBS | HIT KILL | HIT NO KILL | MISS | MISSILE GROUNDED | EXCEEDED RANGE | THRUSTERS |
|---|---|---|---|---|---|---|---|---|
| 1 | | | | | | | | |
| 2 | | | | | | | | |
| 3 | | | | | | | | |
| 4 | | | | | | | | |
| 5 | | | | | | | | |

## FIRING TABLE 6

STATIONARY, MOVING, AND MULTIPLE TARGETS, FAMILIARIZATION ONLY

EVENT A. OPTICAL SIGHT, SITTING.

| TGT | RANGE | *OBS | HIT KILL | HIT NO KILL | MISS | MISSILE GROUNDED | EXCEEDED RANGE | THRUSTERS |
|---|---|---|---|---|---|---|---|---|
| 1 | | | | | | | | |
| 2 | | | | | | | | |
| 3 | | | | | | | | |
| 4 | | | | | | | | |
| 5 | | | | | | | | |

EVENT B. OPTICAL SIGHT, STANDING.

| TGT | RANGE | *OBS | HIT KILL | HIT NO KILL | MISS | MISSILE GROUNDED | EXCEEDED RANGE | THRUSTERS |
|---|---|---|---|---|---|---|---|---|
| 1 | | | | | | | | |
| 2 | | | | | | | | |
| 3 | | | | | | | | |
| 4 | | | | | | | | |
| 5 | | | | | | | | |

EVENT C. THERMAL SIGHT, SITTING.

| TGT | RANGE | *OBS | HIT KILL | HIT NO KILL | MISS | MISSILE GROUNDED | EXCEEDED RANGE | THRUSTERS |
|---|---|---|---|---|---|---|---|---|
| 1 | | | | | | | | |
| 2 | | | | | | | | |
| 3 | | | | | | | | |
| 4 | | | | | | | | |
| 5 | | | | | | | | |

EVENT D. THERMAL SIGHT, STANDING.

| TGT | RANGE | *OBS | HIT KILL | HIT NO KILL | MISS | MISSILE GROUNDED | EXCEEDED RANGE | THRUSTERS |
|---|---|---|---|---|---|---|---|---|
| 1 | | | | | | | | |
| 2 | | | | | | | | |
| 3 | | | | | | | | |
| 4 | | | | | | | | |
| 5 | | | | | | | | |

REMARKS:

* OBSCURATION

SCORER'S NAME (PRINT) ______________________________

SCORER'S SIGNATURE ______________________________

DA FORM 7446-R, DEC 2001

USAPA V1.00

# DRAGON GUNNERY TRAINER (DGT) SCORECARD

## MULTIPLE TARGETS

For use of this form see FM 3-23.24 (FM 23-24); the proponent agency is TRADOC.

DATA REQUIRED BY PRIVACY ACT OF 1974. AUTHORITY: 10 USC 3012(g)/Executive Order 9397. PRINCIPLE PURPOSE(S): Aids in individual training on targets at various ranges. ROUTINE USE(S): To evaluate individual proficiency. SSN used for positive identification purposes only. MANDATORY OR VOLUNTARY DISCLOSURE AND EFFECT ON INDIVIDUAL NOT PROVIDING INFORMATION: Voluntary. Individuals not providing information cannot be rated or scored on a mass basis.

| LAST NAME | FIRST NAME | MI | SSN | UNIT NO. | POINT NO. | DATE (YYYYMMDD) |
|---|---|---|---|---|---|---|
| | | | | | | |

### FIRING TABLE 7

DGT PRACTICE QUALIFICATION
DISK: DGT #14
PERCENT HIT ZONE SETTING
50 % 75 % 100 %

EVENT A. OPTICAL SIGHT.

| | | | TEST | | | |
|---|---|---|---|---|---|---|
| ENG | MSN | OBSC | GO | NO GO | TS | TH |
| 1 | 1 | 0.5 | | | | |
| 2 | 2 | 1.5 | | | | |
| 3 | 7 | 0.5 | | | | |
| 4 | 8 | 1.5 | | | | |
| 5 | 9 | 0.5 | | | | |
| 6 | 10 | 1.5 | | | | |
| 7 | 11 | 0.5 | | | | |
| 8 | 12 | 1.5 | | | | |
| 9 | 13 | 0.5 | | | | |
| 10 | 14 | 1.5 | | | | |

EVENT B. THERMAL SIGHT.

| | | | TEST | | | |
|---|---|---|---|---|---|---|
| ENG | MSN | OBSC | GO | NO GO | TS | TH |
| 1 | 26 | 0.0 | | | | |
| 2 | 27 | 0.0 | | | | |
| 3 | 28 | 0.0 | | | | |
| 4 | 29 | 0.0 | | | | |
| 5 | 30 | 0.0 | | | | |
| 6 | 26 | 0.0 | | | | |
| 7 | 27 | 0.0 | | | | |
| 8 | 28 | 0.0 | | | | |
| 9 | 29 | 0.0 | | | | |
| 10 | 30 | 0.0 | | | | |

### FIRING TABLE 8

DGT QUALIFICATION
DISK: DGT #15
PERCENT HIT ZONE SETTING
50 % 75 % 100 %

EVENT A. OPTICAL SIGHT.

| | | | TEST | | | | RETEST | | | | RETEST | | | |
|---|---|---|---|---|---|---|---|---|---|---|---|---|---|---|
| ENG | MSN | OBSC | GO | NO GO | TS | TH | GO | NO GO | TS | TH | GO | NO GO | TS | TH |
| 1 | 1 | 0.5 | | | | | | | | | | | | |
| 2 | 6 | 1.5 | | | | | | | | | | | | |
| 3 | 7 | 0.5 | | | | | | | | | | | | |
| 4 | 4 | 0.0 | | | | | | | | | | | | |
| 5 | 9 | 0.5 | | | | | | | | | | | | |
| 6 | 11 | 1.5 | | | | | | | | | | | | |
| 7 | 12 | 0.5 | | | | | | | | | | | | |
| 8 | 13 | 1.5 | | | | | | | | | | | | |
| 9 | 14 | 0.5 | | | | | | | | | | | | |
| 10 | 15 | 1.5 | | | | | | | | | | | | |

EVENT B. THERMAL SIGHT.

| | | | TEST | | | | RETEST | | | | RETEST | | | |
|---|---|---|---|---|---|---|---|---|---|---|---|---|---|---|
| ENG | MSN | OBSC | GO | NO GO | TS | TH | GO | NO GO | TS | TH | GO | NO GO | TS | TH |
| 1 | 17 | 0.0 | | | | | | | | | | | | |
| 2 | 18 | 0.0 | | | | | | | | | | | | |
| 3 | 19 | 0.0 | | | | | | | | | | | | |
| 4 | 20 | 0.0 | | | | | | | | | | | | |
| 5 | 21 | 0.0 | | | | | | | | | | | | |
| 6 | 22 | 0.0 | | | | | | | | | | | | |
| 7 | 23 | 0.0 | | | | | | | | | | | | |
| 8 | 24 | 0.0 | | | | | | | | | | | | |
| 9 | 25 | 0.0 | | | | | | | | | | | | |
| 10 | 26 | 0.0 | | | | | | | | | | | | |

LEGEND: ENG - ENGAGEMENT
MSN - MISSION
OBSC - OBSCURATION
TH - THRUSTERS (NARROW COLUMN)
TS - TRACKING SCORE (NARROW COLUMN)

REMARKS:

**SCORING (CHECK ONE):**

| | | |
|---|---|---|
| EXPERT | ☐ | 19 to 20 |
| SHARPSHOOTER | ☐ | 17 to 18 |
| MARKSMAN | ☐ | 16 |
| UNQUALIFIED | ☐ | Less than 16 |

SCORER'S NAME (PRINT) ______________________ SCORER'S SIGNATURE ______________________

**DA FORM 7447-R, DEC 2001** USAPA V1.00

Made in the USA
Monee, IL
05 November 2026

67808314R00157